Praise for *Acquainted*

"As you read these pages, your life will ch[...] of it will change. The night we're all fan[...] thing, deeper, fuller, older, younger, more [...] more spectacular and, yes, more magic[...] —**Margaret Atwood,** *Globe and Mail* (Toronto)

"An enjoyable and instructive read."—*Boston Globe*

"Intriguing . . . a portrait of darkness in all its forms . . . The writer's poetic side is no barrier to his account of the various sciences of the night. Whether he is writing about circadian rhythms or the physics of sunsets it is done with a light hand, but an illuminating one . . . It makes for a perfect bedtime read."—*Economist*

"No better guide can be imagined. A poet as well as essayist, Dewdney laces his account with snatches of poetry and scientific observation, giving pleasure and instruction in equal measure. Take the book to bed. Better yet, take it out onto the porch or patio. Read it until the twilight fades and the first stars twinkle. Then lay it aside and enjoy the spectacle."—*Dallas Morning News*

"Lovely . . . Dewdney's restless intelligence comes at his subject from all directions . . . Heady stuff, and the Toronto-based Dewdney knits it all together with elegant language, reportorial excursions and personal anecdotes . . . Fascinating."—*Time Out New York*

"Judging by his books, Dewdney would make a splendid dinner companion, capable of conversing on just about any topic with genuine interest, and with a warehouse of fascinating facts and observations at his command."—*The Onion*

"Weaving history with mythology, cosmology, and biology, Dewdney has created a mosaic . . . of musings that will no doubt delight night owls as well as those who prefer to spend the dark hours snoring."—*Discover* magazine

"A delightful compendium that charts the nocturnal phases—planetary, human and animal—of life . . . Tautly written in a highly condensed yet readable voice, this tour of the manifold nocturnal realm is a superbly meticulous feat."—*Publishers Weekly* (**starred review**)

"A magical tour of night's great landmarks . . . A eulogistic and very personal treatment of a world to itself, full of incident and lovely as a Whistler nocturne."—*Kirkus Reviews* (**starred review**)

ACQUAINTED WITH THE NIGHT

*Excursions Through the
World After Dark*

CHRISTOPHER DEWDNEY

BLOOMSBURY

For Calla and Tristan

Published by Bloomsbury Publishing, New York and London
Distributed to the trade by Holtzbrinck Publishers

All papers used by Bloomsbury Publishing are natural, recyclable products made from wood grown in well-managed forests. The manufacturing processes conform to the environmental regulations of the country of origin.

The Library of Congress has cataloged the hardcover edition as follows:
Dewdney, Christopher, 1951–
Acquainted with the night: excursions through the world after dark / Christopher Dewdney.
p. cm.
ISBN 1–58234–396–9 (hc)
1. Dewdney, Christopher, 1951– 2. Poets, Canadian—20th century—Biography. 3. Night in literature. 4. Night. I. Title.

PR9199.3.D48Z46 2004
811'.54–dc22
2003023167

First published in the U.S. by Bloomsbury Publishing in 2004
This paperback edition published in 2005

Paperback ISBN 1-58234-599-6
ISBN-13 9781582345994

1 3 5 7 9 10 8 6 4 2

Typeset by Hewer Text Ltd, Edinburgh
Printed in the United States of America by
Quebecor World Fairfield

CONTENTS

1

FIRST NIGHT

I have been one acquainted with the night
I have walked out in rain—and back in rain.
I have outwalked the furthest city light.

ROBERT FROST

I LOVE NIGHT. SOME of my earliest memories are of magical summer evenings, the excitement I felt at night's arrival, its dark splendor. Later, when I was eleven, there were hot summer nights, especially if the moon was bright, when I felt irresistibly drawn outside. I'd wait until my parents were asleep and then sneak out of the house, avoiding the creaky parts of the wooden stairs and the oak floors in the hallway. After quietly shutting the back door behind me, I was free, deliciously alone in the warm night air. A bolt of pure electric joy would rush through me as I stepped into the bright stillness of the moonlit yard.

We lived at the edge of a forest, so I'd hop the rail fence and blend into the trees. Even without moonlight my night vision was good enough to avoid stepping on twigs and dry leaves. Imagining I was a puma or a leopard, I'd walk silently through the forest, a creature free in the North American night. Although I didn't know it at the time, by exercising my night vision I was proving Victor Hugo's maxim "Strange to say, the luminous world is the invisible world; the luminous world is that which we do not see. Our eyes of flesh see only night." I was one with the darkness, the forest and the animals of the night.

In a sense I still am. My fascination with night has continued unabated. I have published poems about night and I still enjoy long night walks, though they pass through city neighborhoods instead of forests. Of course I'm not the only one who is infatuated with night. Melissa, the seven-year-old daughter of a friend, told me (in confidence) that, on certain special nights—she called them "wish nights"—she goes outside and captures some of the twilight in a small wooden box her mother gave her. "That way," she said, "I can look inside the box during the day and see my magic night in there. I can only open the box just a little, though, otherwise all the dark will leak out." Her box reminded me of William Blake's "Crystal Cabinet": "And within it opens into a world / And a lovely little moony night." I don't expect everyone to love night as Melissa and I do, but we probably have all had at least one remarkable night, a night we'll always remember—that first evening date with the love of our lives, a summer night camping under the stars, perhaps a memorable night at the movies as a child. But it seems that darkness is especially sweet for new lovers. For them the evening is never long enough. As Thomas Moore wrote:

> *Fly not yet; 't is just the hour*
> *When pleasure, like the midnight flower*
> *That scorns the eye of vulgar light,*
> *Begins to bloom for sons of night*
> *And maids who love the moon.*

Deeply, intimately, we are shaped by night. It is part of us. The rhythms of our bodies, the ebb and flow of our moods, the very pulse of our minds, are vitally linked to the daily cycle of light and dark. For some, night may be a time of anxiety, of loneliness and apprehension; for others, it is a celebratory time of freedom from work, of sensual pleasures and entertainments. Night is when we can put the worries of the day behind, yet it is also a frontier in which we are blind, where unseen dangers lurk. We are creatures of the day, and night has historically been our adversary, an exception to the rule of light. But in the larger universe the reverse is true, night is the rule and light is the exception. Daylight is bracketed by darkness, and our sun floats within the immensity of an endless cosmic night. But this universal night is not barren: it is the fertile void from which all things, even light itself, are born.

Night is profoundly in our souls and minds, our hearts and bodies. It is woven into our language. There are a thousand and one Arabian nights and each night has a thousand eyes. There is music in the night and a nightingale sings in Berkeley Square. Night crawlers glisten on residential lawns, while downtown, night owls rub shoulders with fly-by-nights. The Victorians slept in nightshirts by the glow of their oil-burning night-lights. We used to don nightcaps for bed—now we down a couple before calling it a night. Soldiers use night vision on midnight patrols, as night porters dream of nightclubs where night-hawks party into the wee hours. There are night watchmen, night-walkers, and night stalkers. Ladies of the night come and go, speaking of Caravaggio. And finally, there is the night before the night of nights.

Every landscape, every geography, every place on our planet has its own unique flavor of darkness. But even within that local flavor, each night can be different from the night before, even if sometimes it seems the same night can replay for weeks. There are summer nights when the air is charged with erotic anticipation. There are cool autumn nights when vacant patios are lit by halogen floodlights— empty theaters waiting for a scene that will never be staged. There are deep nights when the planets gather in the sky like moons and the twilight is as blue as a Thai sapphire. In the Arctic there are nights that last for months, nights when the dark hours are so cold that rubber tires freeze solid, become brittle enough to shatter with a single blow from a hammer. There are Caribbean nights above shining palm trees that glitter in the moonlight, their shadows punctuating the smooth white beach like asterisks—the air fragrant with jasmine. And there are city nights inhabited by pacing insomniacs, where streetlights shine through wind-restless trees to cast sliding mosaic shadows on the pavement. There are cold April nights in the Swedish countryside, where small streams form geometric fingers of ice at their edges.

But what is night? For the American poet Robinson Jeffers, night was a kind of luminous darkness, the calm center of beauty: "the splendor without rays, the shining shadow, / Peace-bringer, the matrix of all shining and quieter of shining." When something is as common and familiar as night is, it becomes hard to describe, in fact to even discern, what its exact qualities are. The *Oxford English Dictionary* defines it simply as "The period of darkness which intervenes between day and day, that part of the natural day during which no light is received from the sun, the time between evening and morning." When I started researching this subject, I asked myself, what could I possibly discover about night that most of my readers don't already know? What could I say about it that everyone hasn't experienced? After all, night is much more than commonplace: its occurrence is completely ubiquitous, predictable, and inevitable. But as I began to gather information, to read what novelists and poets and scientists had written about night, I soon realized that it was a huge, labyrinthine topic.

What started as a trickle of leads and connections soon turned into a torrent of information—more and more aspects of night that I hadn't considered, or even known about, revealed themselves to me. Books, music, magazines, and movies began to pile up in my study, and as file folders (both physical and digital) became stuffed and overflowing, I realized that the culture of night was extraordinarily rich with mythology, romance, medicine, superstition, natural history, art, magic, festivals, dreams, music, science, psychology, and even economics.

This book is a record of my exploration of night, my excitement and sometimes my amazement at what night portends, how it moves and shapes us. I was surprised, for example, to find out that the most fundamental aspect of night, its darkness, isn't at all straightforward. The fact that the entire night sky *wasn't* as bright as the sun itself was an enigma only solved by science in the 1920s, well after Einstein had written his general theory of relativity. I also learned that sleep isn't necessarily a healing period; in truth it is sometimes quite the opposite. I discovered that there are vampire-like parrots called keas that prey on sheep in the New Zealand dusk. I became familiar with the U.S. Naval Observatory's three stages of twilight. I talked to firework technicians and city councilors, to cab drivers and sex trade workers, to police officers and naturalists. I became a student of the night. *Acquainted with the Night* is a compendium of night, a wide-ranging investigation of all things nocturnal. But at the same time, I hope that it transcends a mere inventory of night's chattels and becomes something more—a tour of the great landmarks of night, an excursion into the exotic, magical, sometimes terrifying, sometimes exhilarating realm of darkness.

Part of the process of writing about night was trying to gain a perspective that made it new, not commonplace. The exercise I found most effective was to imagine that I was explaining night to beings from another planet that had two opposed suns and no night at all. For them the entire concept of periodic darkness would be novel. Certainly they would be familiar with shade, with caves and windowless rooms, but they would find the idea of total darkness covering everything—mountains, deserts, cities, and forests—inconceivable. Using this as a starting point, I found it easier to imagine that I was such a stranger, and that I too was exploring night for the first time.

It is this perspective—of recasting the ordinary anew—that I have tried to achieve with this book. The chapters are arranged around a "typical" twelve-hour night, anywhere in the world (though naturally influenced by the fact that I live in Toronto), starting with sunset and ending at dawn. This universal night combines everything nocturnal into a single representative night. Each chapter corresponds to an hour of the night and introduces what typically happens during that hour of darkness; sunset, twilight, astronomical darkness, midnight, etc. The deeper content of each chapter is also linked to the normal sequence of a typical night, though some of the things that go on after dark are not always restricted to a specific hour of the evening.

Naming Night

Night, the beloved. Night, when words fade and things come alive. When the destructive analysis of day is done, and all that is truly important becomes whole and sound again. When man reassembles his fragmentary self and grows with the calm of a tree.

ANTOINE DE SAINT-EXUPÉRY

Every word we use has a history. Some words change over time, while others change very little. The English word *night* hasn't changed in the last four hundred years, although it did change a great deal before then. English belongs to the Indo-European language group, which has one of the oldest linguistic heritages on the planet. Many English words can be traced back thousands of years to their origin in Mesopotamia, and the word *night* is one of these. Although our modern term seems to have little in common with its ancient synonyms, the association of night with the sound of the letter *n* was already present more than four thousand years ago. It was there, for example, in the ancestral Hittite term for night, *nekus*, and is also recorded in the old Sanskrit term for darkness, *nakti*. It was also there in the Egyptian name of the sky goddess, Nut, who was often depicted with stars on her body.

In the first few centuries A.D. the Sanskrit term for night made its way to northern Europe where it joined the vocabulary of the developing Germanic languages. Among these was the newly emerging English language. By A.D. 825 the word for night, *Nacht*, was the same in German and English. German has maintained the term unchanged up to the present, but in English it kept changing as the English language was influenced by the vocabularies of other European languages. By the late ninth century *nacht* had changed to *niht* and in the mid tenth century it became *næht*. Around 1200 it transformed to *nahht* and more than two hundred years later, in the mid fifteenth century, it took on the form of *nyght*, which held until the mid sixteenth century, when it became *nycht*. Less than fifty years later, just after the Renaissance, when William Shakespeare wrote *A Midsummer Night's Dream*, the spelling of the word reached its modern form: *night*.

Linguistically, and mythically, night has almost always been personified as a woman. In genderized languages *night* is invariably feminine: in French it is *la nuit*, in Italian *la notte*, and in Spanish *la noche*. Although many associations with night are masculine, the global consensus seems to be that night is female. Night is subtler and more receptive than the day and in this sense, according to some traditional lores, it is deeply feminine and mysterious. The ancient Egyptians regarded the night sky as female and in the world's mythologies the moon is almost universally considered female. Night is personified by a goddess in the sacred Hindu texts of the Rig-Veda, and the Taoist religion also views the night as female, as an endless well of yin energy and darkness, which, in turn, breeds yang energy and light. Because Taoists regard darkness as the source of everything in the universe, twilight also represents the apotheosis of enlightenment. According to *The Wisdom of Laotse*:

He who is conscious of the bright,
But keeps to the dark,
Becomes the model for the world.
He has the eternal power which never errs,
And returns again to the Primordial Nothingness.

Poets have sung the praises of night for thousands of years, and they too have invariably personified night as female. In his "Hymn to Night," Romantic poet Henry Wadsworth Longfellow described night as his lover:

I heard the trailing garments of the Night
Sweep through her marble halls!
I saw her sable skirts all fringed with light
From the celestial walls!
I felt her presence, by its spell of might,
Stoop o'er me from above;
The calm, majestic presence of the Night,
As of the one I love.

And Percy Bysshe Shelly, another Romantic, also described night as a walking woman, "Like the lamps of the air when Night walks forth," while Australian poet Frank S. Williamson, in a sort of reverse personification, described his lover as if she were the night:

SHE comes as comes the summer night,
Violet, perfumed, clad with stars,
To heal the eyes hurt by the light
Flung by Day's brandish'd scimitars.
The parted crimson of her lips
Like sunset clouds that slowly die
When twilight with cool finger-tips
Unbraids her tresses in the sky.

We will encounter this feminine aspect of night again and again: in nocturnal celebrations, in music, art, films, and mythology. Fertility is dark and female, and even though the creation stories of monotheism have masculine creators, the earliest creation myths feature feminine deities of night.

Opening Night:
The Origins of Night in Mythology and Science

In Genesis, the first book of the Hebrew Bible, God names the darkness "night" after creating day:

> " 'Let there be light;' and there was light. God saw that the light was good, and God separated the light from the darkness. God called the light Day, and the darkness He called Night."

The Genesis account is similar to other religious and mythological explanations of the origin of night, but very few others have such an abstract and cosmological account of night's origin as the Bible's. In most cosmogonies darkness predates everything, and creation has to be wrested from it. Unlike the Genesis account, many of the world's other origin myths often assume that the earth and sun preceded creation. At most they explain the lack of sun at night, such as the ancient Egyptian belief that the sun passed through the body of the sky goddess, Nut, each night, but in general they assumed that night and day existed a priori.

The ancient Greek account of the beginning of night, though personified in a similar way to the Egyptian myths, is closer to the Bible's account than it is to the Egyptian explanation—even if, deistically speaking, the Greek version is much more carnal. In Hesiod's *Theogeny*, written in the seventh century B.C., there is a description of how night began: "First of all, the Void came into being, next broad-bosomed Earth, the solid and eternal home of all, and Eros (Desire), the most beautiful of the immortal gods, who in every man and every god softens the sinews and overpowers the prudent purpose of the mind. Out of Void came Darkness and black Night, and out of Night came Light and Day, her children conceived after union in love with Darkness." Aristophanes, the Greek comic poet of the fifth century B.C., described a slightly different origin for Eros in his play *The Birds*. In this version the goddess Night (also known as Nyx) produces an egg from which Eros hatches: "in the infinite bosom of Erebus, Night with black wings first produced an egg without a seed. From it, in the course of the seasons, Eros was born—the desired, whose back sparkled with golden wings, Eros like swift whirlwinds." The goddess Night, it seems, was able to conceive both immaculately and carnally, at least according to the ancient Greeks.

The Maori of New Zealand had a similar cosmogony. They believed that the sky and earth were personified as two original gods, Rangi and Papatu-a-nuka, respectively. The rest of the Maori's seventy gods were the children of Rangi and Papatu-a-nuka. According to the Maori legend, in the beginning there was no light, only darkness, much like the Genesis account, and Hesiod's. But because the sky pressed down on the earth, there also was no space for the seventy gods. After considering slaying their parents, the seventy gods began to push the sky apart from the earth physically, standing on one another's shoulders to stretch the sky even higher. It is then that the first light appeared, and accordingly, the first night. The Maori genesis myth, with its emphasis on an almost mechanical explanation for darkness, foreshadows the most recent explanation of night's origin, the scientific genesis.

We humans are a practical lot, and science is one of the most practical tools we have yet devised. It has given us an account of creation that, although less lyrical than the King James Version of

Genesis, is no less dramatic. A scientific genesis would go something like this: In the beginning there was nothing—no universe, no space, and no time—less than a vacuum. Since there was less than nothing, there was nothing even to compare nothingness with—an inconceivable state. But in this nothingness a unique event occurred: an infinitesimal blip, maybe an intimation, a tiny precognitive echo of the immensity to follow.

Astrophysicists call this extraordinary event—this miracle of something happening within an infinite eternity of nothingness—a "vacuum fluctuation." In turn, the vacuum fluctuation created what the same astrophysicists refer to as the primal "singularity." A singularity is an extraordinarily dense point of mass and energy that defies the general laws of physics. A black hole, for example, is a type of singularity. From that very first singularity, which was about the size of a dime, the entire universe blossomed in a tremendous explosion called the big bang and is still expanding today. That was 14.5 billion years ago.

For the first few hundred thousand years of its existence, our universe consisted of glowing plasma—there was no darkness, only light. Then, as matter began forming out of the plasma, gravity appeared, and soon dark, unlit patches of space began to proliferate within the cooling plasma. After that, the light in the universe went out completely and the universe entered a period of absolute darkness. This first (but not last) cosmic night lasted five hundred million years.

But within the darkness the universe was evolving, gravity began exerting its force, and the once-glowing plasma was slowly transforming into clouds of gas. Inside this utter blackness the embryos of stars and galaxies began to form. Eventually, at some point about fourteen billion years ago, the nuclear fires of an original star ignited and the first starlight shone in the darkness. More and more stars began to catch fire and eventually they formed the first galaxies. Gradually the dark space between these stars and galaxies became the largest part of the continuously expanding universe and provided the darkness necessary for a night sky. So the very first night in the universe occurred then, a little less than fourteen billion years ago, on a barren planet or asteroid orbiting the youngest star there ever was.

But what about the first night on earth? When did that occur? According to the scientific genesis, our own solar system was born approximately 4.6 billion years ago in a vast nebula of dust and gas afloat in the interstellar void. Perhaps initiated by the shock wave of a nearby exploding star, a section of the nebula began to collapse on itself in a runaway gravitational attraction. This nebula was incredibly massive, and as billions of tons of matter were condensed at its core, it began to spin from centrifugal force. Then, as it had happened billions of times already in the past ten billion years, at the densest center of the cloud of dust and gas a red glow appeared, the beginning of a new star, our sun. Out of darkness, out of compressed interstellar dust, came light, and out of the disc of dust and gases spinning around the new sun came the planets.

So when was the first night on earth? Was the establishment of night and day a gradual process or an abrupt one? We can answer the first question roughly by looking at the history of our planet. The earth coalesced into its current shape, a slightly oblate spheroid, about 4.5 billion years ago, so the primary night had to have occurred sometime close to that, within, say, a few million years. Because night needs daylight to define it, night could only have started when sunlight reached the surface of our planet, after the dust of creation had been scrubbed from earth's sky. This was probably a gradual process and possibly so incremental that it would be hard to pinpoint the first night exactly. (Perhaps, though, it might have been a faster-paced event. A recent lunar theory proposed by Japanese astrophysicist Eiichiro Kokubo suggests that the moon's original dust cloud was cleared away within a month of the moon's formation. It's possible the same thing happened on earth, and perhaps there really was an initial sunset after which the stars shone in a clear night sky for the first time.)

What we do know is that the first night wasn't very long. The earth was spinning much faster then, before the moon put the brakes on. The gravitational interaction of the earth and the moon slows the earth's rotation by about two milliseconds each century, twenty seconds per million years, and two hours every three hundred and eighty million years. About nine hundred million years ago, for example, there were four hundred and eighty-one eighteen-hour days in a year, and the average night was only nine hours long. (Interestingly enough, these estimates have been verified by paleontology. Three-hundred-and-seventy-million-year-old fossil corals have four hundred daily growth rings in one yearly cycle, exactly the number of days predicted by the formula!)

If you could travel back in time a billion years, the earth would appear very different. There would be rivers and mountains and deserts, but no land plants. Aquatic life was well established by then, though the creatures that were large enough to be seen would look so strange that you'd think you were on an alien planet. At night, looking up at the stars, you'd see none of today's constellations. That might be somewhat disconcerting, but even more disconcerting would be the speed of the moonrise. A visibly larger moon (it was once much closer to the earth) would seem to leap up from the horizon, and as it crossed the night sky, its pace would be noticeably faster; in fact, it would appear to literally sail through the stars. But to experience the first night on our planet, we'd have to travel even farther back in time, much farther, billions of years before the beginning of life.

If we extend the same formula for calculating the length of the day back to the very beginning of the earth, 4.5 billion years ago, it shows the first night (after the dust had been cleared from the earth's sky) would have lasted only minutes! The stars would appear to wheel dizzily through the heavens and the moon, if there were one (according to a recent theory the moon was created by a large asteroid impact a few million years after the formation of the earth), would careen through the stars like a cannonball. The centrifugal force of such a rapidly spinning earth would be enough to make you light-footed, but you probably wouldn't notice it, because the red-hot, molten surface and toxic atmosphere would require asbestos snow-shoes and a flame-retardant spacesuit.

Nights are thankfully much more serene now, and the pace of the twelve-hour night described in this book seems languorous in comparison with those first frenetic evenings. Looking billions of years into the future of our solar system, physicists predict that the pace of the earth's spin will eventually slow until it stops, and night will be frozen on one side of the planet. It will be a very cold night on a very inhospitable earth. But it is unlikely humans will be around, at least in their present form, to take in the spectacle. For interstellar night beckons us, and we have already flown through the deep night of space to the moon.

Day unto day uttereth speech, and night unto night showeth knowledge.

PSALM 19:2

As I write this, it is a hot August night, a Saturday. My study window is open to let in some cool air, but stray sounds are also coming in. I've stopped writing because a noisy row has broken out in a neighbor's yard. I can't see what's going on, but it sounds like raccoons fighting. They're making eerie caterwauling cries and sinister chuckling noises. I can hear the leaves of trees moving in the wind. From a party in someone else's yard come the sounds of glasses tinkling and laughter and talking. It's a typical summer evening. I realize now, with some smugness, that in terms of research I have it easy, my subject comes to me. No matter where I am or what I do, night will come.

Night is infinite, and a clear night like tonight is a window onto limitless space. Looking out my study window, I can see stars that are millions of light-years away. By day the sky is opaque with blue, or it languishes under an even lower ceiling of clouds, but the night sky opens up into space, it stretches out forever. The darkness both conceals and liberates us, and even a cloudy night has the tinge of eternity. Night gives us permission to hope, to wish, to dream, to be whomever we wish. And beyond the city lights there is a wild night in the country, where fragrant foliage frames the long, dazzled streak of a rising moon reflected on water.

Irreducibly, night is intimate. When our sense of sight is diminished, we become more private; when our range of action is limited by darkness, we retreat indoors and often into ourselves. With night comes personal time, time alone for many and a lonely time for a few. Evening can be a time for self-evaluation, and for anticipating the next day from the contemplative vantage of the night before. Night is a harbor for sleep and dreaming, and in our homes night is spent in beds and bedrooms. Night is also anonymous, and we sometimes venture by night what we are afraid of doing by day. Nighttime is a romantic realm of love and lovers—moonlight touches more passion than does sunlight.

Yet, for all our nocturnal privacy and seclusion, we humans are curious creatures and we wonder about the dark world around us. There is always someone, somewhere, looking up into the night sky and wondering. Sometimes we look through telescopes, sometimes we simply pause briefly to take in the grandeur of a beautiful evening. The night sky has told us more about our place within the cosmos than has the day. Some of the most profound discoveries of science and physics were hidden like riddles among the stars, waiting for patient explorers and their telescopes to discover them. It is from the night that we learned our planet was one among many other planets orbiting a star among millions of other stars in a galaxy among millions of galaxies. It is from the night we learned about black holes and antimatter, about the origins of the universe itself. As Galileo wrote in 1623: "Philosophy is written in this grand book—I mean the universe—which stands continually open to our gaze, but it cannot be understood unless one first learns to comprehend the language and interpret the characters in which it is written. It is written in the language of mathematics, and its characters are triangles, circles, and other geometrical figures, without which it is humanly impossible to understand a single word of it; without these, one is wandering about in a dark labyrinth."

If white is the presence of all colors then black is the potential of all colors. Only in the darkness of night do neon signs dazzle, northern lights dance, fireflies pulse, stars glitter, phosphorescent plankton glow, and fireworks astonish. The night bends down, it speaks to us in silent languages. All hope and purity are in its depths. Night is a collective planetary spectacle, it is a mysterious, magical realm, and it is a frontier that we are still exploring.

2

THE GARDENS OF THE HESPERIDES: SUNSET—6 P.M.

I T IS SIX O'CLOCK in the evening of our universal day, with its twelve hours of sunshine and twelve hours of night. The heat of the afternoon is beginning to diminish and the light breezes that played through the open windows a few hours ago have become more occasional. The shadows of buildings and trees are lengthening, and as they stretch and intermingle, they begin to form the foundation of a twilight that will eventually rise and blend with the deeper twilight of night.

Behind the buildings and trees the sun is sinking rapidly and its red disc is already touching the horizon. A brilliantly colored sunset is in progress. Above the horizon floats a mountain range of purple clouds that seem motionless, though more peaks appear to be forming, gathering out of a sky that has turned from blue to a pale aquamarine. This is the first hour of twilight, and the beginning of night. Sunset, the gateway to night, has begun.

> How the old mountains drip with sunset,
> And the brake of dun!
> How the hemlocks are tipped in tinsel
> By the wizard sun!
>
> Then, how the fire ebbs like billows,
> Touching all the grass
> With a departing, sapphire feature,
> As if a duchess pass!
>
> EMILY DICKINSON

In the world's cities, at least those on the leading edge of twilight, the streetlights are coming on: incandescent gas inside the giant, frosted bulbs flickering first orange, then blue, violet, and finally, bright white. It's rush hour and the highways and streets are filled with cars, and even in parks the air is blue with exhaust. From the perspective of a nearby apartment balcony, the gridlocked commuters on the expressway make it look like a segmented, metallic snake—though one that's barely inching along. Impatience and hunger, combined with low blood sugar, add up to a volatile mix that fuels anger. Statistically, the greatest number of road-rage incidents occur around this time. The lights of houses and apartment buildings are starting to come on, competing with the golden reflection of the sunset in the windows of the condominium buildings to the east. Power consumption is at its highest and the hydroelectric grid is close to peak capacity. Waiters are arriving at restaurants to begin preparations for the dinner hour, glasses are polished, tables set, and wine uncorked. The bars are beginning to fill with office employees enjoying a drink after work.

In the residential sections of town, dinner is being prepared and televisions wink on, one after the other, as the news hour begins. The light is fading rapidly in the country also and the first few crickets are starting to call, though on hot summer evenings the cicadas will still be buzzing in the trees for some time after sunset. Singly, and in flocks, birds are on the wing, silently flying low and fast to their night roosts. The country day shift is winding down, and during this twilit hour the transition between diurnal and nocturnal animals will take place. Some nocturnal creatures are unwelcome—mosquitoes are most active for the first hour just after sunset. But there are a host of night fliers that feed on them, and soon they'll be on the wing also.

Sunset Spectacular

Who has not seen in imagination, when looking into the sunset sky, the gardens
of the Hesperides, and the foundation of all those fables?

HENRY DAVID THOREAU

A sensational sunset—and we have all probably seen at least one—is a visual extravaganza. It is monumental, almost grandiose: the sculpted, alpine landscapes of clouds, the intense, saturated colors, and the ever-changing hues that always seem to outdo themselves. Great sunsets build, and a classic sunset ends in a flourish of fluorescent red clouds that set the sky on fire.

The most consistently brilliant sunsets I've seen have been over water, and the best of those were off the west coast of Saint Lucia, in the Leeward Antilles. Many of them rivaled fireworks, the colors were that brilliant. The Saint Lucian sunsets consistently filled half the dome of the sky with atmospheric masterworks. Often they were so complex they were like universes of detail that might take a thousand years to explore. Some evenings the sky resembled a vast surrealist hallucination drenched with pigment, while on other nights the clouds seemed to explode into flames directly overhead. They were utterly glorious, celestial pyrotechnics. If that weren't enough, the Saint Lucian sunsets were almost invariably accented with whimsical, sometimes unbelievable touches: a trio of small electric-orange clouds shaped like waves, for instance, or a gridlike archipelago of fluorescent crimson cloudlets set against a pale, lost blue sky above a moody mountain landscape of lavender clouds.

These Saint Lucian sunsets always started the same way: long, horizontal bands of frothy, mother-of-pearl-colored clouds would build above the western horizon when the sun sank toward the ocean. As the sunset progressed, these bands would light up like pink flames against the opalescent blue sky. Floating under the bands, just above the ocean, were armadas of purple cumulus clouds, sailing with the trade wind, their movement contrasting with the stationary cloud banks above them. Then, as the sunset began to peak, the pink edges of the banded clouds would turn cerise, then red, and the red would deepen, like wildfire, into a lava filigree of incandescent lacework. Many of these features were, of course, lit from beneath, like most sunsets, though the effect in Saint Lucia was more like looking up at the immense, stained-glass dome of a cosmic cathedral. The sky immediately above and behind these scintillating chromatics was an impossible shade of pastel lilac, and closer to the horizon the clouds ignited with evocative reds and oranges that were strangely reminiscent, at least for me, of childhood memories. The Saint Lucian sunset, like all epic sunsets, constantly transformed, gradually yet quickly enough that I had to keep scanning the whole panorama to see all the changing parts. Each sunset was a feast that left me dazzled, awestruck, and visually gorged.

Following every glorious Saint Lucian sunset, like dessert, was a low-level acrobatics display. On the hill sloping down toward the ocean from my sunset-watching deck vantage, in the green-blue Antillean twilight, dozens of narrow-winged bats would skim low over the lawn, wheeling and dipping to within inches of the grass in a silent airborne carnival of frenzied insect-catching. The first time I saw this fleet of sleek dark shapes cartwheeling above the lawn, I thought they were swallows, but later, after one flew close enough for me to get a good look, I realized they were bats, foraging in the twilight. After such fanfare, after the sunset and the bats, night's ultimate arrival seemed almost anticlimactic, though there were deeper mysteries and more subtle spectacles to come in the tropical darkness.

The Sundowners

The sun was just setting behind the edge of a wooded hill, so rich a sunset as would never have ended but for some reason unknown to men, and to be marked

with brighter colors than ordinary in the scroll of time. Though the shadows of
the hills were beginning to steal over the stream, the whole river valley
undulated with mild light, purer and more memorable than the noon.

HENRY DAVID THOREAU

Sunset watching, wherever there is an elevated vista or a westward expanse of open seashore, is a perennial social activity. In some places it is a regular evening ritual, particularly in the summer, when people gather during fair weather to watch the play of colors and cloud formations. Artists and writers have described sunsets since classical times; J.M.W. Turner's famous paintings of sunsets over the canals of Venice come to mind, as does T. S. Eliot's description of the sunset as a "patient etherised upon a table," a strangely apt metaphor that, at the time his poem was published, was regarded as anything but apt. (Shocked reviewers of the early twenties said this metaphor was too ugly and modern to be poetic.)

It speaks to our aesthetic impulse that even the most inartistic among us still enjoys a lovely sunset, though this capability may be deep in our genes, in fact, for we are not the only creatures to enjoy a sunset. There is evidence that we share this appreciation with our closest cousins, the great apes. In an article in *Scientific American*, chimpanzee researcher Adriaan Kortlandt recounted an extraordinary behavior that she observed one evening: "Once I saw a chimpanzee gaze at a particularly beautiful sunset for a full 15 minutes, watching the changing colors [and then] retire to the forest without picking a pawpaw for supper."

Practically every inhabited geographic location on earth claims to have the best sunsets in the world, and individual estimations of what constitutes a good sunset are highly variable. Unlike the Richter scale for earthquakes, there is no rating system for sunsets. I interviewed almost a hundred widely traveled people and asked them if there was anywhere in the world they had been that consistently had great sunsets. The following top-ten list is the result:

1. Lesser Antilles, Caribbean
2. Amboseli National Park, Kenya
3. West coast of Ireland
4. East shores of Lakes Huron and Superior, Canada
5. West coast of Zanzibar, Tanzania
6. Arizona
7. Thailand (near Phuket)
8. Greek islands
9. Uruguay
10. Bali

My list, of course, is not absolutely definitive, but I gathered enough anecdotal evidence and overlapping testimonials to elect certain places as sunset capitals. (I was gratified to see, for instance, that the Antillean sunsets were universally appreciated.) Sunsets were evaluated according to range and intensity of color, the fantasticalness of cloud formations, dependability (good sunsets, day after day), size (how much of the sky they took up), and duration (how long they lasted). It is easy to see from the list that great sunsets almost always occur over large bodies of water and are best viewed from the western coast of whatever land overlooks the water.

The Physics of Sunset

O'er me, like a regal tent,
Cloudy-ribbed, the sunset bent,
Purple-curtained, fringed with gold,
Looped in many a wind-swung fold
JOHN GREENLEAF WHITTIER

The colors of the sunset that have inspired so much poetry and art are usually dominated by reds, yellows, and oranges. This is because the light from the sun slices obliquely through the air instead of shining straight down from above. Light from the setting sun passes through about thirty-eight times more atmosphere than light from overhead sun. Because air molecules selectively deflect blue and violet, the increase in the density of air between the setting sun and us screens out the blue end of the spectrum. The more air that light travels through, the redder that light will appear, which is why the setting sun and the rising moon appear orange or reddish. This is also why the western sky at sunset is saturated with the red end of the spectrum.

Airborne dust particles also affect the color of the sun, and they enhance and augment the effects of the atmosphere. After the volcanic explosion of Krakatoa in the South Pacific in 1883, the largest volcanic eruption in recorded history, sunsets around the world were affected. Fine ash from the volcano, blown fifty kilometers into the stratosphere by the force of the explosion, quickly circled the globe. Even three months after the explosion of Krakatoa some of the sunsets in North America were so brilliant that on one occasion fire engines were dispatched in New York and New Haven to what were thought to be huge conflagrations in the western parts of those cities. These spectacular sunsets continued for almost three years after the eruption.

But the atmosphere has even more tricks up its voluminous sleeve for our twilit delectation. Airborne particles and density certainly create a spectacular palette of effects, but there are other, more subtle chromatic phenomena that, although fleeting and rare, reward the patient sunset-viewer with an extraordinary display.

The Green Flash

Nobody of any real culture, for instance, ever talks nowadays about the beauty of sunset. Sunsets are quite old fashioned . . . . To admire them is a distinct sign of provincialism of temperament.

OSCAR WILDE

Toronto is cold and miserable in February. After two months of snow and blizzards I need a reminder that summer still exists somewhere, that I don't *have* to hunch through slush in the dreary freezing winter. So, for the last few years I've been spending a few weeks every mid February in Bonaire, one of the "ABC" islands (the others are Aruba and Curaçao) off the north coast of Venezuela. I spend my time snorkeling the exceptionally rich reefs that surround Bonaire, and because of my marine interests, I usually stay at a small but comfortable dive hotel called Bruce Bowker's Carib Inn. There are lots of other divers to talk to and marine-life tips to pick up. Bruce Bowker himself runs the hotel, and there isn't much that goes on there without him being directly involved, from buying the air conditioners for the rooms to personally supervising dives.

The guests are mostly American scuba divers, though there are sometimes a few Europeans and Canadians, and every late afternoon when the sun nears the horizon, a handful of them gather on a covered patio at the ocean's edge for a ritual I was unprepared for, at least the first time. As the sun began to drop behind the sheer, clean line of the ocean horizon, in the stillness that always fell at that hour during clear evenings, everyone would pause their conversations and lean forward intently, looking at the sinking sun. Just after the last red wedge of sun slipped under the horizon, someone would invariably call out, "There it is!" and someone else would say, "Yup, good one!" "What?" I asked Bruce Bowker, who was sitting beside me. "The green flash!" he replied. In my confusion I had missed it, because, as I found out later, the flash lasts only two or three seconds.

I've seen sunsets with groups of amateur sunset-watchers, the kind that gather at almost every ocean and lakeside in the late afternoon to watch the sun going down, but this group was much more specialized, and I was surprised at the familiarity they had with a phenomenon that I had only read about briefly once in *Scientific American*. My understanding from the article was that the green flash was a very unusual, rarely witnessed occurrence. Yet here in Bonaire, it seemed to happen every evening. These divers were not your average sund-owners, but a professional crew of sunset epicureans that even Oscar Wilde would have approved of. They shunned the normal palette of sunset colors incandescing against a backdrop of blue sky and crimson clouds, no matter how spectacular. They only congregated when the sky was cloudless, and conditions were right for observing the green flash.

So just what is this green flash? The green flash, as its name suggests, is a brief green glow that appears above the place where the sun has just set. It is an atmospheric phenomenon that requires certain conditions: a clear sky, fairly calm winds, and the right humidity. When these conditions are met, and only then, there appears, momentarily, a bright green band of light, slightly less than a third of the diameter of the sun, just above the last edge of the sun as it slips below the horizon. Bright green is a bit of an understatement; electric emerald green is more like it. The green flash really is exceptionally beautiful, as I eventually found out. But what causes it?

Because the green flash is like an addendum to the sunset, its genesis is similar. As the sun nears the horizon, the intervening atmosphere deflects most of the blue end of the spectrum, leaving only yellow, orange, and red. But another factor, horizontal banding, caused by atmospheric refraction, is necessary to explain the green flash. Most of us have seen the setting sun (or rising moon), when it is just above the horizon, as flattened. This is because atmospheric refraction distorts the image of the setting sun, widening it and, sometimes, breaking it up into several bands. Not only that, but when the atmosphere between us and the sun gets thick enough, it acts just like a prism, dividing the image of the sun into horizontal bands of color that match the order of the spectrum: starting with red at the bottom and moving up through orange, yellow, green, and finally blue. If conditions are right, then as the sun sets, its image will refract through all those bands, starting with yellow and ending with a bright, emerald green "flash" that hovers briefly above the horizon.

But what about the blue band? As it turns out, with the right conditions, a blue flash, much rarer than the green flash, will very occasionally follow the green one. The blue flash has been described as an intense, shimmering jewel of blue violet that lasts only a second. Naturally, it is the holy grail of flashes, and only a handful of observers can claim to have seen it.

The green flash was woven into the screenplay of Eric Rohmer's award-winning 1986 film *Le rayon vert* (French for "the green flash"), a slow-paced, diaristic film about a young woman from Paris named Delphine who is at loose ends during her summer vacation to the seaside. (It was released under the title *Summer* in English-speaking countries. Rohmer's film was loosely based on Jules Verne's novel of the same name, the final chapter of which also uses the green flash as an image.) Delphine, a shy and slightly melancholy loner, makes no friends during her vacation. On her last day at the resort town Delphine meets a young man at the train station. She decides to spend the afternoon with him and they go for a walk on the boardwalk, where they discover a souvenir store called Le Rayon Vert. They begin to fall in love and end up sitting on a secluded hillside to watch the sun set over the ocean. After the sun sets, they both see the green flash of the movie's title. According to Gallic folklore, witnessing the green flash helps you understand your inner feelings, and it is understood that these two new friends will become lovers. *Le rayon vert* took top prize in the 1986 Venice Film Festival, outdoing *Room with a View* and *Round Midnight*. Perhaps Oscar Wilde was right about conventional sunsets, but I'm not sure he knew about the green flash—if he did I'm sure he would have approved.

The Dawn of Night

I love the night. I love to feel the tide of darkness rising slowly and slowly washing, turning over and over, lifting, floating, all that lies strewn upon the dark beach, all that lies hid in rocky hollows. I love, I love this strange feeling of drifting . . .

KATHERINE MANSFIELD

Night comes when you least expect it. You are making dinner or working late, you look out the window and the sky is already dark. The arrival of night can be elusive, mysterious, and in the city we often don't see it, though we always know when it has fallen. In the country night takes its time. A glorious sunset might flag its approach, yet it seems we can never pinpoint its exact arrival. Nightfall is a subtle process.

As children we have all probably waited for the first star so that we could make our wish. I can remember that the evening sky always seemed too bright to reveal any stars at first, though suddenly, as if it were always there (and it was), the first star would appear: "Star light, star bright, first star I see tonight. I wish I may, I wish I might, have the wish I wish tonight." Still, even with the first stars twinkling, I could see that it wasn't exactly night. So when does night begin? Does it wait in dark corners until twilight and then leak out of closets and basements as the sun sets, like sequestered freedom fighters joining an army of liberation? Does it bide its time in cupboards and caves until planetary night descends and then rise to blend with all the earthly shades and darknesses?

According to the U.S. Naval Observatory there are three stages of twilight: civil twilight, which arrives shortly after sunset and which marks the time when car headlights should be switched on; nautical twilight, which arrives half an hour after civil twilight when it is dark enough that the brightest stars are visible for navigation purposes; and astronomical twilight, which starts more than an hour after sunset when even the faintest stars are visible. Within the first hour after sundown, the first two stages of twilight are attained and watching their progress is a profoundly subtle experience.

> *She comes! She comes!*
> *The sable throne behold*
> *Of Night primaeval, and of*
> *Chaos old!*
> ALEXANDER POPE

There is a park, just east of downtown Toronto, that is shaped like a big, flat-bottomed bowl. It has baseball diamonds on the bottom, and its edges are grassy slopes that form a natural amphitheater. I often meet there with a few friends at sunset during the summer, especially during spells of hot, clear weather. We usually congregate on one high embankment that faces southeast to watch night begin. It was my passion for night that got this group started, but I think my friends enjoy the sublime panorama of night's arrival on their own. In a world of sunset-watchers, I have started my own little cadre of night-watchers who look in the opposite direction from the sun-downers, away from the slow-motion pyrotechnics of the sunset. We enjoy a subtler spectacle, which, if the conditions are right, can be an almost mystical experience. Night-rise.

Certainly there are other circumstances in which to watch night-rise. I've found that airplanes, particularly eastbound flights that take off during the late afternoon and fly into the night, have excellent views of night-rise. Winter evenings are also good, though in the north it's often too cold to enjoy them. But as far as I'm concerned, clear summer evenings are the best, even if the atmosphere is slightly hazy. It is important when watching night-rise to have a wide vista, unimpeded by trees or buildings, opening onto the southeast, like our slope in the park. From this vantage we can see the purple and blue dawn of night, like an anti-dawn, growing deeper and higher as the darkness approaches. If the moon is full, it will rise from this violet sea, orange and impossibly large.

The arrival of night is a compendium of gradual change that has a mesmerizing effect on our little group. The light patter of conversation usually stops just after sunset as we begin to privately admire the subtle colors of night-rise. Though elusive they really are quite lovely—ranging from pastel lavenders and pinks higher in the sky to a deep purple just above the horizon. As the eastern sky turns a deep blue, then deeper blue still, we begin to see what looks like an indefinite, looming dark cloud rising up above the southeastern horizon. The only reason we know it's not a cloud is the fact that the sky is clear, and besides, none of us can make out a definite edge to this dark, violet shape. It is the shadow, or penumbra, of the earth and it curves downward very slightly at each side. As the penumbra deepens and rises, a few ghostly silver stratus clouds often appear, just above the horizon in front of it.

This part of night-rise, the deepening and rising of the penumbra and the end of civil twilight, lasts perhaps ten, possibly fifteen minutes at most, at least at our latitude, because once the earth's shadow angles up into the sky, night-rise speeds up. The darkness in the southeast that seemed like a cloud becomes less definite and soon the sky above us is also darkening quickly. Then, mysteriously, the first stars appear and nautical twilight has begun. Looking behind us, at the sunset, we can see only a glow now. The stars are suddenly appearing everywhere, and we begin our conversations again as we turn to leave the park. In less than thirty minutes we have passed from civil twilight into nautical twilight, and although astronomical twilight is still half an hour away, it seems, at least for us, that night has begun.

Slowly the night blooms, unfurling
Flowers of darkness, covering
The trellised sky, becoming
A bouquet of blackness

FRANK MARSHALL DAVIS

Tonight's Forecast

Does night provide its own weather? Is there a climate specific to night that has its own logic and progression? Although I've always been a bit of an amateur meteorologist, I decided I'd better consult a professional about the facts of nocturnal weather, so I called Harold Hosein, who is the weatherman on a local TV station and has a populist, whimsical style to his forecasts.

I met Harold on a cold November night at the TV studio where he works. He fit our interview in between two weather reports: one for radio and another for TV. Harold Hosein has a genial, informal TV persona, but in person he seemed even friendlier. After I checked in with security, he accompanied me to his corner studio in the newsroom—a small, self-contained television set dominated by a central, large-screen plasma display of Southern Ontario with a live radar sweep and readouts from Environment Canada. He had a desk and a stool, and several more TV screens tuned into weather networks rested on every available surface.

My first question concerned the effect of darkness on temperature; I knew that it got cooler at night but I didn't know by how much. Harold answered that the temperature usually begins dropping at sunset and continues to fall until it reaches a low point, generally between 4 and 7 A.M. "On a clear night in the summer," he went on, "the temperature sinks rapidly after dark and drops about seven degrees Celsius [or about 13 degrees Fahrenheit] over the course of the night. During the winter the temperature can drop as much as ten to twelve degrees Celsius [18 to 22 degrees Fahrenheit] on a clear night." The reason he specified clear nights, he told me, is because on a clear night heat is lost by radiation directly into space, but on a cloudy night this heat is trapped by the cloud layer. As a result, the temperature drops more slowly. "In fact," he continued, "on cloudy summer nights the temperature may not drop at all."

I then asked him about the wind: does that change at night? Harold replied that it depended on where that night occurred. "If you're beside a big lake or an ocean," he said, "the water temperature affects the wind at night. Any prevailing wind usually drops off in the evening, but if there is a large body of water nearby, then the wind will shift at night—from an onshore breeze during the day to an offshore breeze at night." He said this was because the nighttime temperature usually drops below the water temperature, meaning that the relatively warm air above the water will create an updraft that in turn will draw in air from the land. Otherwise, at least in flat regions away from large bodies of water, there are usually no convection winds at night. But in mountainous regions, he went on, very high winds will spring up during the middle of the night, quite suddenly, even on a completely still evening. This is because as the air cools at night, it begins to sink, and in the mountains it collects on the slopes like rain on a roof, gaining mass and momentum as it descends until, by the time it reaches a valley, it is a raging torrent of cool air. These alpine night winds can be quite strong, and very unexpected, like avalanches of cool air. Finally I asked him if humidity levels were affected by darkness. He said that humidity is usually unaffected by night, though on clear nights in the summer there is a small increase in humidity, starting around midnight, as the dew begins to form.

When we finished, I thanked him for his time and stood back and watched the first part of his live weather broadcast. His switch to his

TV persona was unnoticeable, a reason, perhaps, for his popularity. I could see our reflections in the windows that overlooked the night view of the city behind his corner studio, and for a moment I was aware of watching myself watching Harold Hosein, who was himself being watched by thousands through the eye of the camera at the heart of the electric circus of the night metropolis.

The Size and Speed of Night

Night, in astronomical terms, is the dark side of a rotating planetary body in orbit around a star. Only if that planet is spinning, as ours is, will viewers on its surface experience alternating periods of light and dark. Astronomers call the line demarcating shadow from light the "terminator line," and sunset (and dawn) marks the passing of the planetary terminator line over us.

How big is night? The surface area of the earth is about 197,000,000 square miles, of which exactly half, or 98,500,000 square miles, is in shade and covered by darkness at any given time. If we take night proper to mean nautical twilight, as occurs about half an hour after sunset, and deduct it from that part of the earth in shadow, then night covers a surface area of 90,292,000 square miles, or roughly the size of the Pacific Ocean.

How wide is night? Again, by determining the average circumference of the earth (because the earth, being an oblate spheroid, is slightly wider at the equator) as being 24,881 miles, and cutting that in half, we arrive at the width of the dark part of the planet, approximately 12,440 miles. Assuming that night proper starts about half an hour after sunset and deducting that from the total, night is 11,400 miles wide. Looked at another way, the same night stretches from Perth, Australia, to Rio de Janeiro, Brazil.

I ask of thee, beloved Night—
Swift be thine approaching flight,
Come soon, soon!

PERCY BYSSHE SHELLEY

Sometimes when I lie in bed at night, I entertain myself with a little fantasy. I try to physically feel the velocity of the earth's spin. As an exercise it is pretty futile; it amounts to an idle abstraction, because I have no direct sense of our tremendous speed, there are no signs—the wind doesn't howl past my bedroom at hundreds of miles an hour—nor can I "feel" the surface speed of the earth. Certainly there is some evidence that we are spinning; the moon rises and sets, the stars drift across the night sky. But, overall, the actual speed of our planet's spin is another one of those facts, like the distance to the sun, that I have to take on faith. Still, I do try to imagine what it would feel like to be aware of the earth's motion. In fact, whenever a new residence permits, I prefer to have my bed aligned along the direction of the earth's spin, with my head pointing east and my body "streamlined" into the spin of the earth.

As we sleep, we whistle through the darkness, dreaming on the surface of an immense, spinning top. Above us the stars wheel on uncharted voyages to remote worlds while the night carries us on a journey halfway around the globe. But how fast are we moving? The answer lies in latitudes and circumferences. At the equator the circumference of the earth is 40,070 kilometers or 24,899 miles. Because the earth revolves once every twenty-four hours, we can calculate the surface speed of the spinning earth at the equator by dividing the circumference by twenty-four hours, which is around 1,670 kilometers per hour, or 1,040 miles per hour. We don't all move at the same speed because, by latitude, the circumference of the earth decreases toward the poles. In Los Angeles, at 34°N, the surface speed of the earth is 1,384 kilometers per hour, or 860 miles per hour.

Farther north, at 45°N, the latitude of Milan, Bordeaux, Halifax, and Minneapolis, the earth's rotational speed is even slower, 1,180 kilometers per hour, or 733 miles per hour. Rotational speed decreases to 0 miles per hour at each pole, which, combined with the low angle of the sun, is why sunsets last longer at higher latitudes and why above the Arctic Circle sunsets can last for weeks. (Though, if you factor in that the earth orbits the sun at some 67,100 miles per hour, even the poles are whistling through space.) In contrast to the poles, the speed of night is fastest at the equator, where the tropical night falls quickly, "like a curtain," as British colonialists were fond of intoning.

Another mental exercise that gives you a sense of how fast the earth is spinning is to imagine you're standing at some arbitrary location on the earth's surface, and then calculate the speed of the planetary spin for your location using our formula. Imagine, for example, that you're in Los Angeles standing around with a group of friends, perhaps in the parking lot of Griffith Observatory, where scenes from movies as diverse as *Rebel Without a Cause* and *Terminator* were filmed. Then imagine that, by some magic force of will, you could suddenly stop moving while letting the earth's rotating surface slide beneath you. Several things would happen instantaneously. As far as your friends were concerned, you would simply disappear. This is because, relative to anyone standing beside you, you would accelerate to 869 miles per hour (the "spin speed" for Los Angeles) in an instant. Even if you could survive the crushing G-forces of such an abrupt acceleration, you would do well to be above such surface features as the observatory, trees, and hills—they would lurch toward you at 869 miles an hour atop a horizon-wide conveyer belt suddenly moving at lightning speed. The only evidence that your friends would have that you existed would be the sonic boom your body made as it broke the sound barrier over the Pacific Ocean.

But, aside from such wild conceptual exercises, many of us do possess both an abstract notion and a partially witnessed impression of the world's spin—one more placid and perhaps majestic than my imaginary, frantic translation of actual relative velocities. It is particularly at night, with the gentle transit of stars and moon, that the immense rolling of the earth can be sensed. Thomas Hardy, a British novelist and poet, wrote about this sensation in his famous novel *Far from the Madding Crowd*, which, in turn, became the possible inspiration for the Beatles' song "Fool on the Hill."

To persons standing alone on a hill during a clear midnight such as this, the roll of the world eastward is almost a palpable movement. The sensation may be caused by the panoramic glide of the stars past earthly objects, which is perceptible in a few minutes of stillness, or by the better outlook upon space that a hill affords, or by the wind, or by the solitude; but whatever be its origin the impression of riding along is vivid and abiding. The poetry of motion is a phrase much in use, and to enjoy the epic form of that gratification it is necessary to stand on a hill at a small hour of the night, and, having first expanded with a sense of difference from the mass of civilized mankind, who are dreamwrapt and disregardful of all such proceedings at this time, lone and quietly watch your stately progress through the stars.

Time Travel, the International Date Line, and the Progress of Night Across the World

In yet another mental exercise, imagine being able to fly from New York to Paris in a very fast jet, much faster than the Concord, that could get you to Paris in five minutes. If it was, say, 7 P.M. on a Tuesday night in October in New York when you took off, then, when you arrived in Paris a few minutes later, it would be 1:05 A.M. the next day, Wednesday. Technically, in terms of the dark side of the planet, you'd still be in the same night that was just under way in New York, it's only that that particular night would be a little more than half over in Paris.

Flying farther east, to Pakistan, night would be getting old by 5:10
A.M. and when you reached central China the sun would be rising on
Wednesday morning. Wednesday would mature normally as you flew
eastward over the Pacific Ocean until about noon. Then an odd thing
would happen. Instead of flying into late Wednesday afternoon and
then Wednesday night (by the time you got back to New York), you
would be in the middle of the Pacific on Tuesday afternoon, almost
seven hours before you left New York!

So what happened? This is all very confusing. Maybe it's a cynical
plot to prevent us from traveling forward (or backward) in time on fast
airplanes, like a variation of the comic story where Superman spun the
earth backward to reverse time and save Lois Lane. Traveling into
the future does seem naïvely possible given the arbitrary abstraction
of the international date line coupled with a spinning planet, and, to a
certain extent, limited time travel to the future is not only possible,
it's done every night. North American stockbrokers use time travel
when they get the jump on local stock exchanges by looking at
tomorrow's trends today. The Nisei stock exchange in Japan opens
almost half a day ahead of the Nasdaq and patterns developing there
can make or lose money in New York later the same day. But we all
know, as common sense dictates, that there can only be one day and
one night at a time on our globe.

So where do new nights come from? Like new days, they start at the
international date line, an artificial "zero time" zone in the middle of
the Pacific where everything chronological begins. The date line
functions like a switch that advances the calendar by one day every
time the earth revolves around to the same point. It's as if the earth is
turning with a clicker and switch attached to it, and every time the
international date line comes around, it clicks in the new day, like the
rotary dial of a plug-in light timer.

It seems that one of the basic imperatives of our species is to
organize our world into quantifiable units, but as a result of that

abstraction we are sometimes left with contradictions and peculiarities. For example, days have to be inserted into leap years in order to make calendars accurate, and territorial politics influence the shape of the international date line. Countries on the same longitude, which, in a logical world should be in the same time zone, exist in completely different time zones! This was most evident during the Y2K celebrations a few years ago. The governments of several Pacific nations close to the international date line realized years before the turn of the millennium that there would be a lucrative profit to be made from Y2K tourists who would travel en masse to whichever Pacific nation would be the first place on earth to officially enter the new millennium. There was a plethora of claimants, ranging from Tongo to New Zealand, but the first officially sanctioned millennial celebration ultimately occurred on Caroline Island, part of the tiny nation of Kiribati.

With great foresight, in 1993, Kiribati applied to have a slight loop inserted into the date line that included all its islands in one time zone. The wily president knew this would be a windfall for millennial tourism in a few years' time and his promise to make the change got him elected. The Greenwich Observatory sanctioned his proposal and thus it was that his little nation profited greatly from the millennial hoopla. On New Year's Eve, 1999, under the balmy tropical stars over Caroline Island, the new millennium began.

> *I saw Eternity the other night,*
> *Like a great ring of pure and endless light,*
> *All calm, as it was bright,*
> *And round beneath it, Time in hours, days, years,*
> *Driv'n by the spheres*
> *Like a vast shadow moved; in which the world*
> *And all her train were hurled.*

HENRY VAUGHAN

Night, then, must also make its first landfall, at least officially, on Kiribati's Caroline Island. After that it slides westward across the Pacific, slipping at the same time into Siberia in the north and New Zealand in the south. Shortly afterward night floods into Australia, as the sugar gliders come out of their lairs. It then surges on and rising as quickly as a dark monsoon tide, it submerges tropical Indonesia and then Japan. As the streetlights begin to come on in Tokyo, night has already started to darken China. After flooding the watershed of the Yangtze River with twilight, the evening spills onward over Burma and Mongolia to India in the south and central Russia in the north, the night stretching long dark fingers ahead of itself to touch the side of each Himalayan mountain. Crossing the Indian Ocean, night falls on Afghanistan and western Russia simultaneously before settling on Iran, darkening the archaeological sites of the first great city-states. As nighttime prayers begin to ululate from the mosque towers in Saudi Arabia and the Mideast, night swoops into eastern Africa. Then it plunges central Africa into darkness along with Europe—the street-lights in Durban coming on at roughly the same hour as those in Milan and Helsinki. This single night also spans two seasons, summer in Stockholm and winter in Cape Town.

When night has entirely covered Europe and the western Sahara, it leaves the west coast of a twilit Africa and sweeps across the Atlantic Ocean, making landfall first on Iceland and then Greenland. Shortly afterward, evening floods the east coast of South America and covers half the great rain forests of Brazil with darkness before first touching North America at Signal Hill on the eastern tip of Newfoundland. It was from Signal Hill on December 12, 1901, that Guglielmo Marconi made the first transatlantic radio communication, into the tide of darkness.

As night pours into the Caribbean, it also covers the west coast of South America and the great eastern cities of North America. Having left South America in total darkness, night has already begun its traverse of the Pacific Ocean—meanwhile it simultaneously spills into Mexico, the Canadian and American Midwest, and Nunavut within the Arctic Circle. Night comes early in the Rocky Mountains, immersing the valleys first while the sun still catches the highest peaks. The last part of the continental landmass to be darkened by night is Alaska, on the west coast of North America. Night then continues to spread over the central Pacific islands, covering French Polynesia, Hawaii, and finally Samoa before it slips over the date line once more.

But we have gotten ahead of ourselves following this great rush of darkness, for we haven't answered the most basic question about night, the very foundation of night, which is why is the night dark?

Olbers' Paradox: Why Night Is Dark

At the end of the first hour of night, nautical twilight is thickening and we are on the verge of astronomical twilight. In less than half an hour it will be completely dark. But what does it mean to say "completely dark"? What does that darkness consist of? Is it the absence of sunlight, or is it an obscuring shade that submerges light in an opaque fog? Is the night dark because outer space is dark, and if so, why is outer space dark? These simple questions have been asked for millennia, but only recently have they been getting answers.

After the great scientific revolution of the late 1700s, particularly after Newton had formulated his laws of gravity and physics had become a mathematically based discipline, scientists very quickly began to understand more about our universe. This was partially due to the availability of better telescopes, but it was also due to new scientific methods. Scientists had made great strides in unraveling the secrets of nature, and with the study of thermodynamics had codified how matter, energy, light, and heat interacted. As always, with every new scientific breakthrough nagging contradictions were revealed, and by the early nineteenth century, scientists knew enough about the universe to realize that the dark night sky completely contradicted their new knowledge of the stars. The cumulative light from hundreds of billions of stars shining from every direction should, even in an infinite, dust-filled universe, light up the night sky as brightly as daylight.

The scientist who became most identified with this enigma was German astronomer Heinrich Wilhelm Olbers. Olbers practiced medicine in Bremen by day and gazed at the stars from his rooftop observatory by night. He rarely slept for more than four hours a night and his specialty was comets. It was Olbers who in 1826 first officially posed the question "Why is the sky dark at night?" His question, which was really a hypothetical query, framed by the lack of any clear scientific explanation, came to be known as "Olbers' paradox." At first it does seem like a naïve question. After all, common sense and simple observation would dictate, as Shakespeare once wittily surmised, "A great cause of the night is lack of the sun." But common sense and intuition can turn out to be impediments on the way to the deeper, often counterintuitive truths of nature. Most hard-won scientific discoveries are based on meticulous observations and experiments that sometimes seem illogical at first, like Galilleo's proof of the equal attraction of gravity on unequal masses that he reportedly demonstrated at the Leaning Tower of Pisa.

Edgar Allan Poe wrote about Olbers' paradox in a prose poem called "Eureka" that he penned in 1848. In "Eureka" Poe suggested that the solution to Olbers' riddle had to do with the age of the universe, and that the reason stars hadn't filled the universe with light was because they hadn't had enough time to do so. He then came up with an additional, and at the time, rather heretical suggestion, namely that starlight hadn't had time to fill the universe because the universe had, at some point in the past, *not existed* and therefore all the light hadn't had enough time to reach us. Poe's speculations were ridiculed and dismissed for almost a century, but as it turned out, he was right on the money, even though he had no scientific evidence to support his speculations. The scientific proof of Poe's hypothesis, and the answer to Olbers' paradox, was attached to a discovery that changed our view of the universe, and our place in it, forever.

It seems that many of the greatest cosmological discoveries, discoveries that have shaken civilizations, came from quiet night owls looking through their telescopes on clear evenings. Olbers, Galileo, Copernicus, and Kepler all created religious, philosophical, and occasionally political furors with the results of their patient observations. (We will hear more from them later.) But it was left for the twentieth century's most famous astronomer, Edwin Hubble, to solve the riddle of Olbers' paradox and while doing so, to discover the ultimate truth about our universe.

Hubble, like many astronomers before him, was a night person. He did most of his stargazing on summer evenings at the great hundred-inch reflecting telescope at Mount Wilson just outside Pasadena, California. For several months in 1929 he had been wrestling with some irritating observational data that didn't seem to make any sense. Astronomers already knew that when a light-emitting celestial body (like a star) was moving toward the earth, its spectrum was "shifted" toward the blue end. If it was moving away from the earth, its spectrum was shifted toward the red end. For years Hubble had been noticing that the farther away a galaxy was from the earth, the more its light was shifted toward the red end of the spectrum, no matter where in the night sky he looked. That was impossible, he reasoned, because it meant that everything was moving away from earth, which would indicate that our planet was the repulsive center of the entire universe. This was almost the exact reverse of Galileo's discovery three centuries earlier, which had displaced earth from the center of the universe. What Hubble was seeing in the distant night sky went against the whole direction of astronomical discovery, which had been increasingly diminishing our place in the cosmos. Perhaps there was something wrong with the spectral optics of the telescope, perhaps he wasn't doing the math properly.

This apparent contradiction plagued Hubble for quite some time until one starry California night, as Hubble sat writing notes in the empty observatory, he remembered a theory expounded two years earlier by Georges Lemaître, a Belgian priest. Lemaître had proposed that our universe had begun with the explosion of a primeval atom billions of years ago. Hubble suddenly realized why the most distant galaxies were moving away from earth in every direction. There was nothing wrong with his observations or the math indicating the apparent fact that the earth was at the center of some cosmic expansion because, in a certain way, it was! But so was everywhere else. Every point in space was the center of expansion because every point in the universe was expanding away from every other point like raisins in rising bread dough! Hubble had discovered that the universe was expanding, and, by extrapolation, the "big bang" that started it all. The Russian physicist George Gamow confirmed Hubble's theory in 1948, though it was a skeptical astronomer by the name of Fred Hoyle who later coined the term "big bang."

So, that solves the enigma of the red shift, but how does an expanding universe solve Olbers' paradox? Well, according to thermodynamic theory, light from all the stars in all the galaxies would eventually find its way to earth, and even with interstellar dust absorbing more than half of the total light, there would still be enough luminosity to create a blazing night sky. It is only because the universe had a beginning and is still expanding that light does not accumulate. The darkness of finite space causes the darkness of night; they are one and the same.

3

A THOUSAND EYES:
NATURE AT NIGHT—7 P.M.

O fathomless cat . . . secret police of the neighborhood.

PABLO NERUDA

IT IS 7 P.M. ON our ideal night. The western sky is lit by a faint glow, though we are almost at the end of what the U.S. Naval Observatory calls "nautical twilight." Many stars are now visible and the air temperature is dropping. Some local evening breezes are stirring, though overall, the evening is quiet and still. Faint shadows linger under parked cars and bushes but in less than half an hour they will blend into the rest of night when astronomical twilight begins.

In the city the rush hour is starting to thin out, and the ice-jams of cars have almost melted. Dinners are being eaten and commuters are arriving at their homes. The enticing redolence of garlic sautéed in butter overrides the scent of stale car exhaust in the downtown restaurant district and couples stroll with linked arms on their way to their favorite bistros. The skyscrapers of the financial section are ablaze with office lights as the first overnight cleaning staff arrives and the last few workaholics finish their reports for the morning. In the harbor a dinner party is under way on a yacht; tomorrow the boat will sail south, but tonight friends and acquaintances are celebrated. The luminous spectacle of the city, viewed from the deck of the boat, is at its peak, and the combined noise of ventilator fans, traffic, subways, and humming transformers blends into a continuous, subdued roar, a "distant humming," as Mark Strand described it in his poem "Night Piece."

In the country, and some residential sections of the city, the night shift of the animal world is starting to rouse itself. In a few trees there are rustlings and the occasional, strange, nonhuman calls. From the dark sky come sporadic high-pitched squeaks and from higher still the insistent call of a nocturnal bird. Even in the largest cities, and certainly in the forests and woodlots outside of the cities, raccoons, bats, owls, nighthawks, and cats are awakening and emerging from their hollows and roosts. In the tropics, where astronomical twilight comes even faster on the heels of nautical twilight, small lizards that defy gravity begin to scurry up the walls of houses. They are geckos, and they arrange themselves around the lights in patient tableaux, awaiting the feast of insects that will soon be attracted out of the darkness.

The ocean night shift is also under way, though more advanced because the water is darker. Within the coastal waters and coral reefs the diurnal fish have all found nooks and overhanging ledges to shelter them overnight. Nocturnal fishes—the big-eyed squirrelfish and the cardinal fishes, the lionfishes and the moray eels—come out of their holes and crannies in the reef to begin foraging. In the Caribbean the bioluminescent flashlight fish, with its two glowing lights located just under its eyes, ascends from the ocean depths, along with a host of nocturnal cephalopods: squid, octopi, and cuttlefish, some of the most intelligent creatures of the sea. The chambered nautilus, a shelled cephalopod that inhabits the deep water off the shores of New Caledonia in the South Pacific, also feeds at night close to the surface. Many other cephalopods are nocturnal and some of them are huge: the sixteen-foot, one-hundred-and-ten-pound giant octopus of the North Pacific, and, just off the coast of South America, the Humboldt squid, up to ten feet long (and able to eviscerate a diver in less than a second), hunts through the black ocean in search of prey. Squid take the prize for the biggest cephalopods, the largest being the giant squid, with some specimens recorded up to sixty-five feet long. In the deep waters near Madagascar the coelacanth—a strange living fossil with fins that are part legs—also rises from the dark, cool depths to forage near the ocean's surface at nightfall. Small, stinging jellyfish seem to prefer to swim at night, and anyone who has swum more than once in the ocean after dark has probably encountered their painful tentacles.

As the light fades and the temperature drops, so does photosynthesis—trees and plants stop producing oxygen, though they still exude carbon dioxide. Plant growth generally slows down after dark, though some flowers remain open all night so that nocturnal insects can pollinate them. In rain forests, spring ponds, and summer woods the songs of amphibians usher in the twilight—the high trill of spring peepers and toads, the elastic twang of bullfrogs, the strange, almost electronic calls of tropical tree frogs. Night's themes are not written by the same composer as the day's. In the dark fields, deserts, oceans, and forests the night creatures, with their sophisticated perceptual equipment, are on the prowl. The curtain of daylight that hid the stars has been drawn and the Milky Way meanders like glowing mist above the trees. The nineteenth-century American naturalist and writer Henry David Thoreau loved nature at night and while he was living alone in a cabin near Walden Pond he often went on nocturnal walks. Sometimes, on summer nights, he would fish in the darkness. In *Walden* he wrote about trolling one night on the pond itself:

> Sometimes, after staying in a village parlor till the family had all retired, I have returned to the woods, and, partly with a view to the next day's dinner, spent the hours of midnight fishing from a boat by moonlight, serenaded by owls and foxes, and hearing, from time to time, the creaking note of some unknown bird close at hand. These experiences were very memorable and valuable to me— anchored in forty feet of water, and twenty or thirty rods from the shore, surrounded sometimes by thousands of small perch and shiners, dimpling the surface with their tails in the moonlight, and communicating by a long flaxen line with mysterious nocturnal fishes which had their dwelling forty feet below,

or sometimes dragging sixty feet of line about the pond as I drifted in the gentle
night breeze, now and then feeling a slight vibration along it, indicative of some
life prowling about its extremity, of dull uncertain blundering purpose there,
and slow to make up its mind. At length you slowly raise, pulling hand over
hand, some horned pout squeaking and squirming in the upper air. It was very
queer, especially on dark nights, when your thoughts had wandered to vast and
cosmogonal themes in other spheres, to feel this faint jerk, which came to
interrupt your dreams and link you with Nature again. It seemed as if I might
next cast my line upward into the air, as well as downward into this element
which was scarcely more dense. Thus I caught two fishes, as it were, with one
hook.

The Eyes of Night

Tiger, Tiger, burning bright
In the forests of the night,
What immortal hand or eye
Could frame thy fearful symmetry?
WILLIAM BLAKE

For nocturnal animals night is not opaque. For them it is more like a
period of diminished brightness, of comfortable illumination. In
order to see in the night with anything like the detail that nocturnal
animals do, we require night-vision goggles, an expensive piece of
electronic optical equipment originally developed for military use.
There are two types of artificial night-vision systems, active and
passive. In the active system, infrared light, invisible to the unaided
eye, illuminates the viewing area with an infrared floodlight and the
goggles merely pick up light of that wavelength. In contrast, passive
night vision is really a light amplification system, and is much closer
to the way that nocturnal animals see. Invented during the Second

World War, passive night vision uses an intensifier tube that converts photons into electrons and then amplifies them. Passive night-vision images are green because green is the easiest color for the human eye to differentiate contrast within. We've all seen the eerie, viridescent news clips of nighttime military maneuvers executed with night vision, not to mention reports from darkened cities under rocket attack.

Being able to see in the dark confers a predatory, sinister advantage. This frightening aspect of night vision was featured in the film *Silence of the Lambs*, when a serial killer equipped with night-vision goggles was watching Jodie Foster at close range as she stumbled through a dark house. People and creatures that cannot see in the dark are at the mercy of those who can. In the end it was her sense of hearing that saved Jodie Foster's character from certain death, and heightened auditory capabilities are also on every nocturnal animal's checklist of must-have equipment. But for most night creatures, the ability to see in the dark is the greatest single advantage they have.

One way of looking at the night is as an optical desert where illumination is as scarce as water is in the Sahara. Like desert animals that have special adaptations to conserve every drop of water, the retinas of nocturnal creatures have to catch every stray photon in order to stay on top of the game. Paradoxically, for those night creatures that rely on vision, darkness is all about light, and their ability to locate and amplify the vaguest glimmers within an economy of luminous scarcity can make the difference between life and death. Night vision requires large pupil diameters and for that reason nocturnal animals have big eyes. The eyes of owls and cats come to mind, but some of the nocturnal lemurs and tarsiers—small, elfin primates with long, human-looking fingers—have the largest eye-to-body ratio of any mammal. Their eyes seem almost alien. In fact, their eyeballs are so large that, like owls', they cannot be moved in their sockets. Tarsiers have to swivel their head to see their prey.

The primates with the keenest nocturnal vision, better than both lemurs and tarsiers, are the South American night monkeys. They are small monkeys, averaging about two pounds, though their long bushy tails and dark fur (which is, apparently, exquisitely soft) make them look larger. Night monkeys have large eyes, and appear similar to tarsiers, with a small triangular snout and delicate mouth. Their long, slender-fingered, human-looking hands are also similar to tarsiers', but there the parallel stops. Night monkeys are social primates with more complex behavior patterns. They forage at night in the jungle, sometimes hooting like owls, which is why they are also referred to as "owl monkeys." Males and females stay in pairs throughout their lives, much like humans, and adolescent night monkeys may remain with the parents for many years, leading to small, closely knit family groups of four or five individuals. Night monkeys move cautiously and silently through the jungle in the dark eating fruits, nuts, and insects. But there is something special about their eyes that make them unique among nocturnal mammals.

It has always been disconcerting for humans to know that while we are blind in the night, some of the larger and most dangerous predators aren't. Most of the great cats—lions, tigers, jaguars, leopards, and panthers—prefer to hunt by night, and their eyes are well adapted to gather more light. The famous naturalist W. J. Holland described the eyes of the great cats rather poetically: "I have studied the eyes of lions and tigers in the dark. The yellowish-green iris shrinks down to a narrow ring. The windows of the eyes have the curtains drawn back wide, so as to let in all the light which the darkness holds within itself. The great orbs then look like globes of crystal, framed in a narrow band of gold, lying on a background of the blackest velvet, while in their pellucid depths, fires, tinged with the warm glow of blood, play and coruscate."

As well as gathering more light, the surfaces of the cat's retinas are covered with a fine layer of tiny, light-receiving cells called rods and cones, much like our own. But, unlike ours, nocturnal animals' retinas have many more rods than cones. This is because cones are best suited for bright light conditions, where they register color and detail, while rods are better in low light conditions, where they detect motion and shapes. Rods are not as sensitive to color as cones, and the high percentage of rods in cats' retinas means that they, and other night dwellers, probably don't see color as well as we do. In addition to admitting more light through the front of their eyes, all night creatures' eyes also have a tapetum lucidum, a mirrorlike layer of cells behind the retina that reflects light that has already passed through the retina back a second time. The tapetum lucidum doubles the amount of available light. It is also what makes cat's eyes glow green in car headlights at night. This reflection is called nightglow. (Human eyes have no nightglow, though it seems that way sometimes in photographs taken with flash cameras.) Night monkeys are the only nocturnal creatures besides tarsiers that have excellent night vision without nightglow. Primatologists point to this feature we share with tarsiers and night monkeys as an indication of our evolutionary kinship, though we have nothing like their ability to see in the night. Their black, crystalline pupils are bottomless wells that swallow even the faintest glimmer of light.

In addition to larger pupils and a tapetum layer, the denizens of twilight have other optical adaptations to darkness. Many nocturnal animals have vertical pupils to protect their light-sensitive eyes during the day. A vertical or slit pupil can close much more quickly and efficiently than a round one. Curiously, most owls retain round pupils, though they have perhaps the most powerful night vision of any nocturnal creature. Leonardo da Vinci's journals (which he wrote backward, in mirror writing, supposedly to dissuade casual snooping), composed between 1452 and 1519, contain one of the first scientific descriptions of owls' eyes. His conjectures about the power of their night vision were quite advanced for the time and were based solely on his extraordinary abilities of observation. He wrote:

> The eyes of all animals have pupils which have the power to increase or diminish of their own accord, according to the greater or lesser light of the sun or other luminary. In birds, however, the difference is greater, and especially with nocturnal birds of the owl species, such as the long-eared, the white and the brown owls; for with these the pupil increases until it almost covers the whole eye, or diminishes to the size of a grain of millet, preserving all the time its round shape. In the horned owl, which is the largest nocturnal bird, the power of vision is so much increased that even in the faintest glimmer of night, which we call darkness, it can see more distinctly than we in the radiance of noon.

The visual acuity of nocturnal creatures is supplemented by their familiarity with their own territories. They rely on kinesthesia, a concrete sense of the physical dimensions and shapes of the landscape around them. Because the same animal often patrols the same territory each night, it builds up a mental map of the trees, branches, houses, and geography around it. It knows where a certain cave is, where the bend in the river is, where the garden gate has a hole gnawed out of it. A night creature outside of its normal territory is a good deal clumsier than on its home turf. But, besides a heightened sense of kinesthesia, darkness has conferred other evolutionary adaptations on night animals. Their sense of touch is much more sensitive and finely tuned than that of diurnal animals—and with whiskers, sensitive noses, and dexterous front limbs they sniff and feel their way through places too dark for even the keenest night vision. Bats, of course, have devised an entirely unique way of negotiating through the dark and finding their prey. They use ultrasonic echolocation. By bouncing high-frequency squeaks off their surroundings, bats can navigate through labyrinthine foliage and pinpoint small moths with radar efficiency. Through evolution, night has bequeathed extraordinary abilities on its children. Through science, the world of darkness that was once the sole domain of such creatures is yielding its secrets to us.

Bats

Twinkle, twinkle, little bat:
How I wonder what you're at?
Up above the world you fly,
Like a teatray in the sky.
THE MAD HATTER IN
LEWIS CARROLL'S
ALICE IN WONDERLAND

Bats are the only mammals capable of true flight. Their wings consist of a thin, leathery membrane stretched between the elongated fingers of their front arms. A smaller, secondary membrane is webbed between the rear legs of many species. Bats range in size from the little bumblebee bats of Thailand that weigh only two grams, to the gigantic flying fox bats of the Indo-Malaysian tropics with their six-foot wingspans. There are more than nine hundred species of bats, a figure that represents almost a quarter of all mammal species on earth, and they are found on every continent except Antarctica. Bats often roost in colonies, and some of the largest are in the southern United States. It is estimated that five million Mexican free-tailed bats have their home in the Carlsbad Caverns, New Mexico, and the spectacle of the swarm of bats exiting the mouth of these caverns at sundown is one of nature's grandest sights.

I was familiar with bats from an early age because I grew up in a house that was perched on the edge of a forested ravine. Tall trees surrounded our back lawn and on summer evenings I would go into the yard at twilight to watch the early bats as they flitted in the air above our lawn. Their appearance was as mysterious and sudden as the first star: abruptly they would be there, flying in long elliptical circles in the airspace over our yard. They were agile and very quick—their flight had a nervous, almost synaptic precision as they caught gnats and mosquitoes. I would practice high-pitched squeaks to try to attract them, or I would throw dead moths up and watch the bats swoop down precisely and effortlessly to snatch these morsels in midair. Bats, I had previously learned from an article I'd read in *National Geographic*, could fly through the whirling blades of a fan, they were that nimble. The same article had high-speed photos that froze the bats' feeding movements. It had pictures of them gracefully sweeping insects into their mouths with their wings and scooping up moths in the membraneous basket formed by the skin stretched between their rear legs. I knew these things, though I could only imagine them—the feeding movements of the bats over my lawn were much too fast to see clearly. As the night deepened they began to disappear into the forest. Others have noticed the same dispersal. In his *Specimen Days*, Walt Whitman recorded his observations on the night of May 21, 1877: "Earlier I had seen many bats, balancing in the luminous twilight, darting their black forms hither and yon over the river; but now they altogether disappear'd."

No other nocturnal animal has garnered as much superstition and as many myths as have bats—they are the focus of some of the most outlandish and fanciful folktales on record. The Kogi Indians of northern Colombia, descendants of the Tairona, associate the vampire bat with menstruation, saying that when a girl begins to bleed she has been "bitten by the vampire." In Mexico bats are considered "old mice"; in North America women fear that bats will get tangled in their hair. In China they are considered good luck and images of bats are often found on Chinese rugs, textiles, and porcelain—on these objects the depictions are somewhat fanciful and the bats seem to have buggy eyes and slightly curly wing tips. As well, bats have an association with things strange, freakish, or unnaturally quick. If someone or something moves fast, we say it goes like a "bat out of hell," or that someone who acts oddly is "batty" or has "bats in the belfry." And if someone can't see, they are "blind as a bat," although bats themselves actually see quite clearly.

If any bat could claim to be the very wings of night it is the flying fox. The heads of these bats actually look like foxes, and among the bats they are perhaps the most photogenic. Near the Indian village of Piliangulam there is a huge, ancient banyan tree that hosts a five-hundred-strong colony of gigantic flying foxes, the largest bats in the world. According to the villagers, the colony has been roosting in the banyan for more than eighty years. They believe that the bats are protected by a god, Muni, and a small shrine dedicated to Muni sits under the tree. There are several other villages in India that also revere the flying foxes, though generally speaking, most Indian fruit farmers, and fruit farmers throughout the tropics for that matter, regard flying foxes as nocturnal pests that raid their crops.

There is a wonderful scene in *Bridge on the River Kwai*, the 1957 war epic directed by David Lean, where a colony of gigantic flying foxes, resting by day in a tall jungle tree, is disturbed by the main characters as they hike through the forest on a hot, sunny afternoon. Lean, who had a marvelous eye for natural phenomena, was told of the colony while on location and he decided to film the bats as they fluttered up from their roosts. He used the scene in the final version of the film, with the actors standing below the roost, staring up, mesmerized by the aerobatics above them, but he focused mostly on the effect of the bats' shadows sliding over the jungle foliage surrounding the actors. The moving shadows of hundreds of giant flying foxes transformed the forest tableau into a bewildering, hallucinogenic whirlpool, a bat-dappled chiascuro that swirled around the entranced actors. The nocturnal essence of bats is so strong that even though this scene was filmed during the day, it looks more as if it were illuminated by the full moon rather than the noonday sunlight.

Strangely enough, for all our ambivalence about bats, they have consistently been the one night creature that we identify with. (It seems to be a peculiarly male identification, though.) William Shakespeare imagined what it would be like to ride a bat through the night sky:

> On the bat's back I do fly
> After summer merrily.
> Merrily, merrily shall I live now,
> Under the blossom that hangs on the bough.

Batman and Dracula come to mind, but even philosophers imagine what it would be like to be a bat. Philosopher Thomas Nagel used a bat as an extended allegory for his speculations on consciousness and the mind-body dilemma. In his *What Is It Like to Be a Bat?* published in 1979, he proposed a thought experiment to support his thesis. He tried to imagine what it would be like to be a bat flying through a forest at night, experiencing the world from its point of view. His attempt was well informed and seemed almost plausible, but his fantasy, he concluded, lacked one important feature: although we can know everything about the bat—we can understand its flight mechanism and how it uses ultrasonic sound to navigate—in the end we can never really know what it is like to "be" a bat.

Neuro-philosopher Daniel C. Dennett disagreed with Nagel completely. In Dennett's book *Consciousness Explained*, in a section titled "What It Is Like to Be a Bat," he wrote: "Nagel claims that no amount of third-person knowledge could tell us what it is like to be a bat, and I flatly deny that claim." Dennett then went on to defend our innate, human ability to imagine and empathize with other beings based on a real, even if partial, knowledge of their experience. He claimed that although bats may not have language and self-consciousness as we know it, we cannot draw hard, all-or-nothing lines between animals and ourselves. The debate goes on, while bats, unaware of their metaphoric potency, casually employ some of the most highly evolved navigation systems in the natural world each night.

> They know I'm something to be caught
> somewhere in the cemetery hanging upside down
> like a misshapen udder.
>
> ANNE SEXTON

Bats have mastered darkness in a way that few other nocturnal animals have. Although they are sighted—some of them are as keen-sighted as owls—they have abandoned night vision for ultrasonic echolocation. Bats emit a regular series of sonar pulses and listen for the echoes that bounce back from objects around them. As they approach an object, these echoes bounce back in less and less time. Using echolocation they can navigate and feed with extreme accuracy. Some kinds of bats have taken this system to a high degree of sophistication. There is a type of fish-eating bat in South and Central America, the greater bulldog bat, that uses echolocation to detect the tiny ripples formed on the surface of the water by the dorsal fins of small fish. If we could hear the ultrasonic sounds of bats, night would be a much noisier, sleepless time for us. Bats are loud, very loud. One of the most common bats in North America, the little brown bat, has a decibel rating of about 110 dB— louder than a smoke detector alarm! Echolocation makes bats one of the most efficient insectivores of the night, and a single bat can catch and eat more than half its body weight in insects in one evening. But insects are not totally defenseless.

Evolution has given the aerial warfare between bats and insects a relatively new twist. Some nocturnal insects—lacewings, praying mantises, and a few species of moths—have developed echolocation detectors and certain moths even produce "jamming" signals that foil the bat's ultrasonic echolation systems. This escalating technology parallels the developments of electronic countermeasures in modern jet warfare, a fact that hasn't escaped the military. Several airborne assault squadrons and electronic countermeasure units have bats as symbols in their regimental badges, and one, the British joint Royal Navy—Royal Air Force Squadron, has gone one further. Its emblem is an arctiid moth, a species that has the most advanced anti-bat-jamming countermeasures of any insect.

Vampire Bats

Of all the bats, none garners more fascination, repulsion, fear, and superstition than the vampire. Because of Bram Stoker's novel *Dracula*, and the subsequent twentieth-century explosion of Hollywood vampire films, the association of bats, blood sucking, and Transylvanian castles almost completely overshadows the fact that there are only three species of bats in Central and South America that actually are vampires. In fact, it was the first zoological reports of vampire bats, given to learned societies in Britain in 1896, that inspired Bram Stoker to fuse the vampire legends of east-central Europe with the natural history of the Latin American vampire bat. Although these real vampires cannot transform into caped counts, they are as mysterious and legendary as their Hollywood cousins.

Of the three species of vampires, the common vampire bat is the best documented, and the only one that feeds exclusively on mammals. It is fairly small, about the size of a common brown bat, and hunts at night for a single meal of blood. I once saw a marvelous documentary that was part of the *Nature* television series about a vampire hunter named Mark Fern. He was, and probably still is, obsessed with all creatures that consume blood—leeches, mosquitoes, and ticks, among many others—but the episode I watched climaxed with his trip to Central America to realize a dream he'd had for years—to be fed on by a vampire bat.

According to his informants there was a small island in the Caribbean just off the coast of Central America where there had been recent reports of vampire bats feeding on humans. He arrived at the island with a zoologist and a film crew, and after setting up special infrared cameras, he placed himself in front of their lenses, sitting in a chair, as human vampire bat bait. As night fell, he sat patiently in the ghostly, slightly ominous infrared light, feigning sleep, his arm trailing over the armrest of the chair so that his fingers almost touched the ground. A shape flittered by, then another. Then a single vampire alighted on the ground beside his chair.

Unlike other bats, vampires are extraordinarily agile on the ground, and this one scuttled up to Fern's hand like a tiny, animated gargoyle. It put its face close to one finger, probably using the infrared sensors in its nose that revealed where the blood flow was closest to the skin, and then, gingerly at first, began to bite the tip of Fern's index finger. Fern said the bite was quite painful, and he withdrew his hand involuntarily. The bat retreated under Fern's chair but, boldly, didn't fly away, and when Fern put his hand back down, it returned and began to lap up the blood, which, due to an anticoagulant contained in the bat's saliva, was now flowing quite freely. The process remained fairly painful. According to Fern the bat used its tongue to explore the fairly sizable wound it had made and several times during the feeding Fern jerked his hand away.

Fern didn't allow the bat to get a complete blood meal, the whole thing was just too much, but if he had, the vampire would have lapped up the free-flowing blood for about twenty minutes, gorging until it had consumed 60 percent of its own body weight. Vampire bats are experts at taking off from the ground, fully loaded, and this one was no exception. It flew into the night with a sizable blood donation from Fern's finger. Vampire bats are on the wing less than two hours on an average night, if they find an animal to feed on, and they return to their roosts to digest their meals. They only take blood from sleeping humans if there are no other animals to prey on. Some vampire bats specialize in birds, though surely none would ever tackle an owl.

The Ears of Night

When cats run home and light is come,
And dew is cold upon the ground,
And the far-off stream is dumb,
And the whirring sail goes round,
And the whirring sail goes round;
Alone and warming his five wits,
The white owl in the belfry sits.

ALFRED TENNYSON

Owls have embodied what is wise and mysterious about night for thousands of years, and they have also represented what is sinister. Even today, outside the safe perimeter of our rational, technological civilization, the sound of a screech owl ululating in the darkness can send a shiver down our spines. I imagine that sound must have been even eerier thousands of years ago. Nevertheless, the first important mythological references to owls are positive and date from ancient Greece. Athena was the goddess of wisdom, according to Greek mythology, and her favorite bird was the owl, in fact, a particular species of little owl, still found in Greece, called *Athene noctua*. (This owl was one of the first animals to be designated as a protected species, in approximately 500 B.C. Large numbers were said to roost in the

Acropolis, which, de facto, became the first bird sanctuary.) Not only did the Greeks believe that owls could see in the darkness because of their "inner light," they also thought owls had influence over military success—Athena, after all, was also a goddess of war. Greek soldiers believed that if an owl was sighted on the eve of battle, they would prevail over their enemies the following day.

The Hellenistic myths, in turn, were incorporated into the pantheon of Roman mythologies and the Roman goddess Minerva became their version of Athena. She remained the goddess of wisdom and in the guise of an owl she also became the symbol of philosophy, a tradition that has continued through to the present. But even though the owl became a philosophical symbol for the Romans, represented in stone carvings and mosaics, the Romans were never as fond of the real thing as the Greeks were. Owls were considered a bad omen in Rome, a superstition that persists in Italy to this day.

Owls have various and spooky mythological associations all over the world. The Kwakiutl Indians, on the rainforest Pacific coast of Canada, believed that owls were the souls of living people, and if you killed an owl, then the person whose soul it embodied would also die. In northern India a folk belief held that if you ate the eyes of an owl, you would acquire perfect night vision, and southern Indians believed that owls were fortune-tellers according to the number of hoots they made: two hoots bringing success in anything being initiated the following day and three hoots indicating that a new daughter-in-law would soon enter the family.

> *Sweet Suffolk Owl, so trimly dight*
> *With feathers, like a lady bright,*
> *Thou sing'st alone, sitting by night,*
> *Te whit! Te whoo! Te whit! To whit!*
>
> THOMAS VAUTOR

Owl's feathers are adapted for a completely silent flight. In most birds the turbulence created by their wings creates a whooshing sound, something you can hear quite plainly when a duck or a goose flies close overhead. Owls, instead, have a specialized, comblike feature on the leading edge of their primary wing feathers that acts like the spoiler on a race car, reducing turbulence to almost nothing and allowing them to fly without making a sound. Most of their prey never know what hit them. Owls also have spectacular night vision, perhaps one of the most acute of all nocturnal animals—with their forward-facing eyes they have a wide field of stereoscopic vision and can judge distances with precision. Their retinas are thickly carpeted with rod cells that are themselves soaked in a chemical called visual purple that translates the vaguest glimmer of light into an electro-chemical signal. All told, owls' vision is a hundred times more sensitive than humans', but their hearing is even more acute.

Owls can hear all the frequencies of sound that we do, but their reception of the higher frequencies is much better than ours, so they can detect the slightest rustlings made by their prey in grass or the leaves of forest undergrowth. As well, their ability to pinpoint the location of sounds in three dimensions is phenomenal. An owl can calibrate down to the millisecond the difference in the arrival of sound between one ear and the other. If the prey is to its right, then the sound arrives at its right ear a millisecond before it hits the left ear—and owls can detect a difference of about thirty millionths of a second between their left and right ears! The owl swivels its head (their necks are famously flexible; they can look directly behind themselves without moving any part of their bodies except their neck) until the sound is arriving at both ears simultaneously. It then knows that its prey is directly in front of it.

The hearing of owls is augmented by another unique feature: the shallow cones of the facial discs that surround their eyes. These circular "dishes" of feathers catch faint sounds and focus them toward their ears. In a sense, when we look at an owl, we are looking at a creature that has managed to overlap sound and sight in one place: they literally have "ear-eyes." Some owls, particularly the barn owl, take the ability to locate prey three dimensionally to an extraordinary level of sophistication. Barn owls are perhaps the strangest looking of all the owls, with their heart-shaped faces and long white legs, but their strangeness runs deeper than appearance. Not only can the barn owl change the shape of its facial discs (using specialized muscles), guiding sounds more precisely into its ears, but its ear openings themselves are asymmetrical. One of their ears is higher than the other. This allows the barn owl not only to pinpoint the location of its prey from left to right, on a horizontal plane, but also, instantaneously, up and down on the vertical plane. It's as if the barn owl has a three-dimensional Cartesian imaging system. Using the complex processing power of its medulla (that part of the brain that is devoted to analyzing sound), which, with approximately a hundred thousand neurons is three times the size of a crow's medulla, the barn owl calculates the exact location of its prey. Unlike the bat and its echolocation system, the owl uses sound amplification to guide it through the darkness of night.

Researchers have filmed barn owls striking mice in total darkness, using infrared light to illuminate their flights. From thirty feet away, simply by hearing the tiny sounds a mouse's claws make on laboratory linoleum, the barn owl will triangulate the position of the mouse, launch into a steep, noiseless glide, and making any last-minute course corrections, strike the mouse with 100 percent accuracy, time after time after time. Amazing, uncanny, and sinister precision. The beak and talons of night.

Of Nighthawks and Whippoorwills

Nighthawks belong to a family of nocturnal birds called goatsuckers, or *Caprimulgae*. Goatsuckers are represented by about 118 species worldwide, and have names like chuck-will's-widows, frogmouths, whippoorwills, oilbirds, and nightjars. They do not suck goats at night, but rather, feed on insects, which they catch on the wing, a little like bats. In fact, the oilbird uses a batlike echolocation system to navigate through the dark caves it nests in. The brown, gray, black, and white mottled coloring of goatsuckers camouflages them well, particularly when they sleep on branches or on the ground. You often won't see them until they fly up suddenly, like a whippoorwill I surprised once that looked, as it silently flew up into the trees, like a large brown moth.

Over the last thirty years I've noted in my natural history journal the first night of the year when I've heard, or even better, seen, nighthawks, when they return to Toronto from their sojourn in Central America. The common North American nighthawk is a migratory species a little larger than a pigeon, with long, pointed wings. It spends the winters in Central America and migrates north in the late spring following the wakes of humid, tropical high-pressure weather cells. I usually see the first nighthawks around the third week of May. It's easy to identify them, particularly if there is still enough light to reveal the white bar on the underside of their wings. Their call is unique, a sort of raspy "pewt-pewt" sound, and their flight style is also quite unusual, a flutter-glide, flutter-glide pattern. They arrive with great fanfare, it seems to me, calling continuously as they fly into town around sunset.

One of the things I like most about nighthawks is that they are a wild species that has adapted to city living quite readily. In this way they are like raccoons, opossums, bats, and crickets. They lay their eggs on flat tarry roofs that resemble the pine barrens they prefer in nature, and they feast on the multitude of insects that are attracted to the city lights at night. Soon after they arrive in the spring, they mark out their mating territories and begin pairing off. In the city the females select a suitable roof while the males perform extraordinary

ariel displays to impress their mates. On warm summer nights they
flutter up to about eighty feet above a prospective female and then
dive straight down, pulling up out of their dive at the last possible
second. The force of the sudden change in direction combined with
the speed they achieve at this point, up to one hundred miles per
hour, causes their wings to vibrate with a characteristic "thrumming"
sound. If you listen for them on June evenings, you can hear this
vibration, which sounds quite unnatural, over downtown lofts and
factories.

The most outlandish-looking goatsuckers are the frogmouths, a
tropical nighthawk that has a disproportionately large head and a
great, wide mouth that gives them their name. They look vaguely like
homely, bewhiskered owls, though their appearance is so freakish that
it stretches the comparison to say so. They often roost in small groups,
and if disturbed, they will stretch and elongate, pointing their beaks
up in the air and becoming rigid. Their camouflage is so impeccable
that you'd think you were hallucinating, "Weren't those lichen-
covered branches looking at me just a second ago?" There are twelve
species of frogmouth, all very similar looking and distributed through
Malaysia, the Philippines, and Australia. Some of them hunt like
owls, swooping down to snatch up insects and spiders from the
ground, and they sometimes hoot like owls also. Australian nights are
often punctuated by the low, booming call of the tawny frogmouth, a
favorite in nocturnal creature displays at zoos around the world. In
South America there is a related, almost identical-looking species of
goatsucker called the potoo.

There are very few other species of purely nocturnal birds. New
Zealand's famous Kiwi is nocturnal and there are two types of parrot
also native to New Zealand that are seminocturnal. One of them is the
world's heaviest parrot, the flightless kakapo, a golden-green bird
that is currently endangered. It is almost completely nocturnal and
has a marvelous, deep, booming call reminiscent of the tawny

frogmouth's. The other, the kea, is a midsize parrot a little larger than a pigeon, also with golden-green feathers and a long, hooked sharp beak. The kea preys, Harpy-like, on sheep at night. One or several will attack a single sheep, perching on its back and slashing at it with their beaks. Although they do strip off the occasional bit of flesh they also lap up the blood of the sheep. Keas are officially regarded as omnivores, feeding opportunists, but this habit of drinking blood also makes them the world's only vampire parrots.

Nocturnal Insects

> Tis placid midnight, stars are keeping
> Their meek and silent course in heaven;
> Save pale recluse, for knowledge seeking,
> All mortal things to sleep are given.
>
> But see! a wandering Night-moth enters,
> Allured by taper gleaming bright;
> Awhile keeps hovering round, then ventures
> On Goethe's mystic page to light.
>
> THOMAS CARLYLE

While butterflies snooze and bumblebees nap under leaves, a host of other insects are on the wing. Moths, beetles, lacewings, and stoneflies take to the air, foraging for food and dodging the hungry bats and birds that hunt them. If there are artificial lights nearby, most nocturnal insects will head straight for them. Why? Well, it's no secret that insects are attracted to ultraviolet light, which is why outdoor lights are often yellow (the opposite of violet), but the main reason is the moon. On clear nights flying insects steer by the moon.

The moon is a good navigation point since it is usually high enough not to be obscured by hills and trees and it is sufficiently far enough away to act as a landmark. This is an effect we've all probably experienced while driving at night in automobiles—the buildings and trees speed by, but the moon seems to keep constant pace with your car, no matter how fast you go. So it is with airborne insects. If they use the moon as a landmark and keep it in the same relative position as they fly, say, to their left or right, then they can travel in a straight path in whatever direction they need to travel. But, because insects are mostly "hardwired" they cannot learn that a streetlight or a porch light is not the moon. When they try to navigate with such an artificial "moon," keeping it in the same position relative to their direction of travel, they have to keep compensating and their flight path curves. As they continue to compensate, their trajectory spirals ever closer to the light until they finally arrive. From the insects' perspective it is as if they have inexplicably flown to the moon. No wonder they simply land and fold their wings back, basking in the warmth and glow of the moon's light. For them there is nowhere left to go. They will stay until the light is turned off, or the sun rises, their lunar hypnosis is that deep.

> soon did night display
> More wonders than it veil'd; innumerous tribes
> From the wood-cover swarm'd, and darkness made
> Their beauties visible.
>
> ROBERT SOUTHEY

There are many insects that we associate with night. Moths come to mind, as do june bugs, which belong to the same family as the sacred scarabs of the ancient Egyptians. Perhaps the most famous nocturnal insect, at least for performing night's soundtrack, is the cricket. Summer nights wouldn't be summery without the sound of crickets, as any Hollywood ambient-sound technician knows. There are many species of crickets, almost four thousand worldwide, and most sing at night. In the tropics there are giant crickets that chirp so loudly they wake up babies and sleepers have to use earplugs to keep out the deafening racket. In the more temperate latitudes, though, the song of crickets, along with that of their relatives, the katydids, is much more soporific.

But the superstars of nocturnal insects would have to be the giant silk moths. Some might say that their status is more cultlike than popular, yet no one could deny their glamour. G. W. Cable wrote of an astonishing encounter with captive-bred saturniid moths:

> When, hypocritically clad in dressing-gown and slippers, I stopped at my guest's inner door and Fontenette opened it just enough to let me in, I saw, indeed, a wonderful sight! The entomologist had lighted up the room, and it was filled, filled! with gorgeous moths as large as my hand and all of a kind, dancing across one another's airy paths in a bewildering maze, or alighting and quivering on this thing and that. The mosquito-net, draping almost from ceiling to floor, was beflowered with them, majestically displaying in splendid alternation their upper and under colours, or, with wings lifted and vibrant, tipping to one side and another as they crept up the white mesh, like painted and gilded sails in a fairies' regatta.

On sultry nights of early hot weather in the northern latitudes, or during the rainy season in the tropics, these large butterfly-size moths come to the lights and, after alighting, open their faintly trembling wings like revelations. The giant silk moths, or saturniid moths as they are known scientifically, always create a stir. In North America they are represented by the spectacular cecropia moth, with its six-inch wingspan that spreads to reveal a Rorschach bliss of red, black, brown, pink, white, and orange. Several times I have seen the improbably lovely luna moth, also of North America, with its pale green wings and its long, undulating wing tails. In Africa there is a giant luna moth called the African moon moth, twice as big and, unbelievably, twice as beautiful as the North American luna moth. In Europe the stunning peacock moth is a regular visitor to night lights in the country, and the immense, drop-dead-gorgeous atlas moth of Indonesia has a wingspan exceeding twelve inches. The atlas moth's wings, with their transparent eyespots and smoky brown accents, have a futuristic, oriental look. As well, the tips of both their forewings extend out into a shape that resembles the profile of a snake's head, and on each of these wing tips is the unmistakable image of a reptilian mouth and eye that seems painted on. Mimicry taken to an almost cosmic dimension.

Certainly most giant silk moths' wings have eyespots, and perhaps these frighten predators, but the lavish designs and subtle colorations seem to have no function besides our delight in them. Giant silk moths only live a few weeks, and, like supermodels, they eat very little. In fact, most of them have no mouths. They live only to copulate and procreate, and they attract each other not by sight, but by smell. The camouflaged wings of giant silk moths are like pure invention, or perhaps the art of an alien civilization— as if perfumes and rivers and smoke had been transferred by some magic process onto them. The night is full of hidden wonders, but the giant saturniid moths that flutter around lights on early summer evenings are the most ravishing incarnation of night. They are like emblematic talismans, all the more mysterious because they themselves are unaware that they depict the essence of darkness. How could night hide its deepest, most secret beauty in blackness? And what is the purpose of being so fabulously beautiful if it's not to attract other moths? Why can't these moths see each other the way we see them? These are some of the deepest enigmas of the night, and the answers are as elusive as the darkness itself.

Pale fireflies pulsed within the meadow-mist
Their halos, wavering thistledowns of light
JAMES RUSSELL LOWELL

Silk moths are not the only insects playing the mating game in the darkness, and some insects, like the firefly, rely almost entirely on vision instead of scent. Anyone who has seen fireflies, or lightning bugs, as they are sometimes called, will tell you how spellbinding they are. They might also recount how, as a child on some enchanted June evening, they had chased fireflies on their grandparents' lawn. Even when you've seen many fireflies, you're still never quite ready for that first glimpse, when in the darkness you see a moving green ember shine briefly and wonder, momentarily, what it is. And even if by the second flash you know that it's a firefly, there's still a simmering, unconscious wonder that lurks underneath your rational knowledge. The first flash of a firefly always brushes me with something unknown, the supernatural.

Once, as a child, I caught a firefly and I was surprised to see how drab it was in the daylight. I was also surprised to see that it was a beetle—fireflies should really be called "firebeetles." There are nearly two thousand species of firefly in the world, the majority of which are located in the tropics, though fireflies can be found throughout the temperate regions as well. Their flashing lights, which are a mating display meant to attract males to females, come in many colors according to species; green, orange, red, white, and blue-green. As well, the rate of flashing varies from species to species. Female fireflies are often flightless and are called glowworms. They clamber up on plants and light up very brightly, attracting male fireflies. In some species the females glow almost continuously.

Although we still haven't completely understood how fireflies produce light, we know that it is a form of bioluminescence, a chemical reaction that takes place in specialized cells. There are many other creatures that use bioluminescence, particularly in the ocean, but none is so familiar as the firefly. Fireflies create bioluminescence by combining three distinct organic chemicals—luciferin, luciferase, and oxygen—in the light-producing organs located at the end of their abdomens. Luciferin is the heat-resistant substrate, luciferase is the catalyst, and oxygen is the fuel. The light produced by fireflies, by any bioluminescent organism in fact, is 98 percent efficient. That is to say 98 percent of the energy that goes into making the light is translated directly into light. The rest is dissipated as heat. Electrical engineers have never been able to match this type of efficiency despite the work of scientists who have tried to reproduce bioluminescence since it was first discovered decades ago. The achievement of "cold light," as it has been called, is still a holy grail of electrical engineering.

> *Among these trees, night by night, did show themselves an infinite swarme of fierie seeming wormes flying in the aire, whose bodies (no bigger than an ordinarie flie) did make a shew, and giue such light as euery twigge on euery tree had beene a lighted candle, or as if that place had beene the starry spheare.*
>
> F. FLETCHER, *Drake's Voyage*

Of all the firefly displays in the world, none rival those in the tropics, where thousands of fireflies will alight on a single, large tree. There are several species that do this. The males and females of one particular species will land by the hundreds and then begin to clamber rapidly up and down the branches, flashing furiously as they do so. The effect, from a distance, is extraordinary, as the whole tree pulses with light and develops, in the rain forest mist, a cool, nebulaic halo. But there is another species, *Photuris pyrolis*, that outdoes all the others. These fireflies select a single, huge tree and then, uncannily, synchronize their thousands of flashes to a single on-off pulse. Zoologists who work in tropical forests at night and who have come upon these trees without warning report that the sight is literally unbelievable at first, the resemblance to a flashing neon sign is so strong!

There is a wonderful, albeit Victorian, description of these marvelous, synchronized fireflies written by Douglas Cameron, in his *Our Tropical Possessions in Malayan India*, published in 1865.

The bushes literally swarm with fireflies, which flash out their intermittent light almost contemporaneously; the effect being that for an instant the exact outline of all the bushes stands prominently forward, as if lit up with electric sparks, and next moment all is jetty dark—darker from the momentary illumination that preceded. These flashes succeed one another every 3 or 4 seconds for about 10 minutes, when an interval of similar duration takes place; as if to allow the insects to regain their electric or phosphoric vigour.

Bioluminescence in the Ocean

I had heard of bioluminescent plankton that glowed in the ocean at night, though my first actual experience with it was on the north coast of Cuba, on another trip to escape the Canadian winter with my daughter. We had arrived at our rooms just at sunset and I insisted that we get into the water anyway to fulfill my fantasy of shoveling snow in the morning and swimming in the Caribbean in the afternoon of the same day, even though our flight had been delayed and the transfers had taken longer than expected. It turned out to be fortunate that we had to wait until dark to get into the ocean, because after we had been swimming for a little while, I noticed that every splash, every movement we made underwater produced tiny green flashes. I thought at first that I was jet-lagged and seeing things, until I realized that these sparks of light were bioluminescent plankton, a phenomenon that I had read about but never witnessed. We quickly took advantage of the situation and discovered that we could stir long, glowing arcs of green sparks in the water if we ran our hands through it with outstretched fingers. Like tiny aquatic fireflies these bioluminescent organisms use the same mechanism to produce light as their land-based cousins.

There are other forms of bioluminescence in the ocean, including the flashlight fish of the Caribbean reefs mentioned earlier in this chapter that use a slightly different method to produce their light. The deep-sea fish of the ocean depths that carry their own navigation lights, like the running lights along tractor trailers on night highways, use yet another. One of the most spectacular bioluminscent organisms is the Bermuda fireworm, which glows as brightly as the firefly. Christopher Columbus's diary includes what is likely the account of the Bermuda fireworms that spawned, and continue to do so today, in great glowing patches of iridescent green just beneath the surface of the ocean in the West Indies. But for pure, opulent magic, Gerald Durrell's encounter with bioluminescence ("phosphorescence," as he calls it) as a young boy in Corfu, Greece, is unsurpassed. In a passage from *My Family and Other Animals*, Durrell recounts a fantastic display of both types of bioluminescence, airborne and oceanic.

Never had we seen so many fireflies congregated in one spot; they flickered through the trees in swarms, they crawled on the grass, the bushes and the olive-trunks, they drifted in swarms over our heads and landed on the rugs, like green embers. Glittering streams of them flew out over the bay, swirling over the water, and then, right on cue, the porpoises appeared, swimming in line into the bay, rocking rhythmically through the water, their backs as if painted with phosphorous. In the center of the bay they swam round, diving and rolling, occasionally leaping high in the air and falling back into a conflagration of light. With the fireflies above and the illuminated porpoises below it was a fantastic sight. We could even see the luminous trails beneath the surface where the porpoises swam in fiery patterns across the sandy bottom, and when they leapt high in the air drops of emerald glowing water flicked from them, and you could not tell if it was phosphorescence or fireflies you were looking at. For an hour or so we watched this pageant, and then slowly the fireflies drifted back inland and farther down the coast. Then the porpoises lined up and sped out to sea, leaving a flaming path behind them that flickered and glowed, and then died slowly, like a glowing branch laid across the bay.

Nocturnal animals animate the night, they fill it with presence and a sort of mysterious splendor—the distant, rasping call of nighthawks through an open bedroom window, the glowing eyes of a family of raccoons caught like momentary rubies in the headlights of a car. Anyone who has heard the song of a nightingale knows the enchanted, lyric music of the night. Although night creatures are mysterious and often associated with superstitions, they provide us with company in the dark hours. It is fascinating, perhaps even comforting, to know that as we settle down into our domestic, nightly routines, they are outside, moving surely through the darkness.

4

THE CHILDREN'S HOUR—8 P.M.

Between the dark and the daylight,
When the night is beginning to lower,
Comes a pause in the day's occupations,
That is known as the Children's Hour.
HENRY WADSWORTH
LONGFELLOW

B Y 8 P.M. THE FAINT afterglow of the sunset in the western sky is gone and astronomical twilight, the final stage of the U.S. Naval Observatory's three stages of twilight, is well established. Wherever astronomical twilight falls in the world, at least below the Arctic Circle, the entire sky is now a deep, indigo blue and even the smallest stars are visible. Night will remain this dark until just before dawn. Rush hour is over, though some new pockets of congestion are starting to form on arterial routes as reverse commuters filter into the entertainment and restaurant districts of the city. The office towers of the financial district are still bright, floating cubes of light in the night sky, but gradually, office by office, their windows are starting to wink out. Downtown, the streets are repopulating with pedestrians and cars.

In the residential sections of the city meals are being finished and dishes have begun to be washed. Windows are shut, blinds drawn and curtains closed. Families, couples, and singles are gathering in front of screens; some log on to the Internet but almost everyone else, at least in the industrialized nations, tunes in to sitcoms, variety shows, and movies on TV. In these countries 8 P.M. marks the beginning of the peak TV-viewing period, otherwise known as prime time, and the glow of the TV screen floods living rooms with constantly shifting pastel light. From the vantage of the street the darkened TV rooms flicker like electric, blue aquariums. For most, the indoor world is shifting into high gear, and the second wind supplied by dinner sustains the routines of the evening in these last few hours before sleep.

Supper ended much earlier in the villages and farms of the country and a few late chores are attended to under the blaze of yard lights. The animals have been herded into barns, though on some farms the cows are left out all night under the stars—dew will form on their fur as they sleep standing in the cool night air. The insistent, empty call of a whippoorwill sounds from a North American woodlot and the field behind the farmhouse shimmers with the songs of crickets. Near community centers in the country and in city parks, stadium lights are switched on, immersing outdoor baseball diamonds, soccer fields, and tennis courts with light. Soon, scores of insects will begin circling the tall floodlights above these nocturnal athletic theaters.

In one home a little girl is alone in her bedroom, holding her favorite doll and methodically putting the rest of her dolls in her closet and then closing the door on them. She doesn't trust them at night. Her favorite doll is an exception; she'll sleep with her. For very young children, night is like a forbidden zone that they rarely experience on their own. Usually they are asleep after dark, and only rarely are they allowed outside at night. There are some occasions on which children experience the darkness: on holidays like Christmas Eve or Halloween, or perhaps a wedding reception or a night car ride, but for the most part, night is an adult time. For that reason the child's night is much more populated with imaginary beings than the adult's night—children's evenings are filled with magic and monsters.

According to recent statistics, young children between the ages of six and ten are going to bed almost half an hour later each decade, though eight o'clock is still a general curfew for most youngsters. By then they are brushing their teeth and putting on their pajamas while their parents look forward to a few free hours of quiet time after household chores—perhaps a glass of wine and a romantic movie— but first they have to make sure that their sons and daughters are secure in their beds and primed for sleep. For that they will use a dependable sedative—the bedtime story.

Night can be an anxious, phobic time for children. First of all, they must spend hours alone, away from their parents and sometimes even their siblings, in the debilitating darkness. Second, since children have no real limits on their emotions, they experience catastrophic extremes of despair and fear without any of the self-protective denial mechanisms that adults have accumulated. As a result, children can suffer acutely from separation anxiety, lone-liness, and sheer terror. And, as we all know from our own child-hoods, a young imagination can wreak havoc. One of the most common childhood nocturnal phobias, probably one that we've all entertained, is the paralyzing fear that a monster lurks under the bed. Invariably this nameless horror will reach out to snatch a dangling foot or hand if one should stray over the edge of the bed. Another common source of anxiety is the closet; perhaps there are monsters in there as well.

Of course parents employ all sorts of strategies to ensure that their children's anxieties are kept at bay. Because children are creatures of habit, particularly around bedtime, the evening rituals of bath and story help establish a calm, stable atmosphere. Young children need company, and teddy bears and dolls become bedmates and friends who accompany them on their journey into darkness and dreams. Yet still, nothing beats the security of sleeping with Mom and Dad, though sometimes parents have to be firm to make sure their matrimonial beds don't become nurseries. So, in the end, making the children's bedroom a cozy, secure place and allaying their fears of the night become priorities in every home with small children, because at night children must navigate between the two nocturnal extremes of wish fulfillment and terror. Also, in our age of electric media, TV shows, books, and movies condition children's ideas of night, and though many of the images they see are negative, many more are positive. And for good reason.

Walt Disney has been the most prolific producer of animated features and cartoons for children in the past few decades, and his films have often portrayed night as an enchanted realm. Disney's images of starry skies and charmed nights, whether it be the "Dance of the Sugarplum Fairies" in *Fantasia*, or fireworks bursting in the evening sky above the Disneyland castle, have romanticized twilight for generations. His depiction of Cinderella's transformation from scullery maid into a princess decked out in occult couture took place at night, and Peter Pan was decidedly a night owl. Tinkerbell was (and remains) strictly nocturnal and Pinocchio received the gift of life from the Blue Fairy after she had descended from a starry sky. Certainly these stories predated Disney, but his visual interpretation of fairy-tale nights was entirely new. Even classic Disney films like *The Absent-Minded Professor* and *Swiss Family Robinson* featured inviting, starry nights.

The film director Steven Spielberg is another nocturnal romantic, and several of his family-viewing epics have had classic night scenes: the image of flying bicycles crossing in front of a full moon became an icon for *E.T., The Extra-Terrestrial*, and his *Close Encounters of the Third Kind* was chock-full of spectacular night scenes. More recently the *Harry Potter* films have reinforced a romantic view of night, and children learn to identify not only with the beauty of night—such as the scene in *Harry Potter and the Philosopher's Stone* where the young students take candlelit boat rides to Hogswart Castle—but also with the sorcery and darker excitements of twilit worlds.

But books, I believe, still have the upper hand when it comes to calming children before sleep, and I have used them as sedatives for my own children. I enjoyed reading storybooks to my daughter and son when they were very young, and no matter how busy I was in the evening, I tried to make time to read them to sleep. I liked these fantasy breaks from reality as much as they did, and sophisticated children's writers aim their books at adults as much as children for that reason. Sometimes I, too, fell asleep after reading the story, so powerful was the soporific effect of bedtime stories. Over the years I read them classics like *Alice in Wonderland* (which, because it takes place in an underground world, is really a sort of twilight tale) and *Paddle to the Sea*, by Holling Clancy Holling.

I also read certain modern children's books about the night, which, aside from familiarizing them with the wider world and stimulating their imagination, aim to reassure children that night, although a mysterious realm, is one that is ultimately safe. Books such as *In the Night Kitchen* and *Goodnight Moon* are set at night and seem to have been written specifically to shape children's experience of sleep, dreaming, and darkness and to make night an inviting, sometimes surreal, but always comforting kingdom.

Victorian Europe and the Birth of Children's Literature

It is only in the last two hundred years that it has become traditional to tuck children in with a fairy tale. According to some sociologists, the idea of childhood, at least as we currently conceive it, was an invention of the Victorian era. Victorians idealized the early years of human development as a special time of life, a time when young imaginations could roam freely without the burden of adult worries. This notion of childhood as a privileged phase of life gained acceptance not only in Victorian England, but also throughout Europe and the newly industrialized world. The proliferation of fairy stories and nursery tales was largely due to increased access to education and rising levels of literacy as well as the expansion of the middle class. A great hunger for juvenile literature soon arose, and it was fed by the first compilations of children's folktales, in particular the Grimm brothers' edition of European fairy tales.

Some of the stories in the Grimm compilation had their origins in the Middle Ages, dating from a time when children's stories were recounted and improvised upon by generations of parents who had memorized these tales from untold repetitions that they themselves had listened to as children. Bedtime stories were recited by candlelight or lamplight to children who often had worked all day (rather than going to school), and some of the stories reflect the harsh reality of their lives. Grimms' treasury of fairy tales represents the inheritance of hundreds of years of folklore, and according to Jungian theorists, it is a treasure trove of clues to the symbolic levels of the unconscious mind. Bedtime stories are not only the forerunners to dreams, they form and shape them by autosuggestion. They are the bridges, the gates to dreams, and the archetypal themes of folkloric parables are buried deep within these tales. As a result, more than a simple story is passed on to our children. Much more.

Although night in these old fairy tales was most often a negative time—a time when, for instance, Hansel and Gretel got lost in the dark forest—night could also be an enchanted time, when wishes came true. Cinderella became a princess with a golden dress at night, and night was when the miller's daughter spun gold out of straw in the story of Rumplestiltskin. It was this positive thread of enchantment and magic that was picked up and perpetuated in the psychologically aware atmosphere of the twentieth century. Night, at least for young imaginations, became a friendlier place.

The Bridge to Dreams: Four Children's Night Classics

One of the first psychologically sophisticated modern books about night was *Goodnight Moon*, published in 1947. It was the product of a collaboration between children's writer Margaret Wise Brown and illustrator Clement Hurd. Brown, the author of another children's book about night called *Wait till the Moon Is Full*, said that she dreamed the story of *Goodnight Moon* one night and that she wrote the book in almost final form upon awakening the following morning.

I first encountered the book as an adult when I read it to my son. *Goodnight Moon* is a seemingly simple story about a young rabbit going to sleep in a large green bedroom. Over the course of the story, the rabbit says good night to various objects in the room, including a pair of mittens, a mouse, and paintings on the wall. The rhyming text is playful, almost humorous, but Clement Hurd's mysterious and magical illustrations are what transform the book from a common children's story to an enigmatic classic.

After having read *Goodnight Moon* several times, I noticed that not only did the story have an almost conspiratorial intensity to it, but that it took place in a timeless, perhaps even cosmic dimension. The "great green room" seemed to exist on some other plane of reality. Certainly the room was accessible and cozy, but it was also surreal. More than a simple room, it is a nocturnally charged space, an existential theater.

In the room are two clocks: one over the fireplace and the other beside the rabbit's bed. At the beginning of the story, the clocks register 7:10 P.M., and each time we turn a page or two, the clocks register ten minutes later. According to the clocks, the whole story lasts an hour, for the rabbit is asleep by 8:10 P.M. The bedroom in *Goodnight Moon* is like a stage on which the unfolding of time affects the room's inhabitants: two kittens play with a ball of wool beside an empty rocking chair on one page, then, on the next page, they sit staring at an old lady who has suddenly appeared in the rocking chair. It seems that things change only when you aren't watching, that time is only present beyond the edge of your vision. Each page is a frozen tableau and somehow this mood captures the essence of night indoors, of being cozy and on the edge of dreams.

The incremental (but implacable) nature of time is shown by another, subtle device in *Goodnight Moon*. As you turn the pages, and then compare them, you see that the room is getting dimmer. Each successive page is a little darker than the previous one, and twilight gathers, gradually diffusing throughout the room despite the cheery circle of light cast by the bedside lamp. Because these pages are dominated by the great green room, it seems as if there is no world outside of it, besides the night sky, that there is only the room, which makes the young rabbit seem as if he exists only in the present. Perhaps this is the source of the spookiness that seems to illuminate the story with a crisp background of neutral apprehension. Maybe there is only this moment, this room. The fire crackling brightly, the kittens playing on the hearth, the blue twilight of the space beyond the dim glow of the rabbit's bed lamp. A lucid, hypnotic feeling pervades everything.

Through the window nearest the bed there is a full moon rising. At one point the rabbit sits up to look at the moon. Because the base of the window obscures the horizon, it seems possible that the room is floating in the night sky, surrounded by the stars on all sides, alone in the immensity of the universe. Finally the little rabbit falls asleep, the room darkens, and the windows grow brighter while the night sky outside turns a pale, moonlit blue. Ultimately, the simple, wry lyricism of *Goodnight Moon* initiated the development of a new kind of children's literature, one that combined poetic imagery with a more subtle, modern narrative.

American critic and poet Randall Jarrell wrote several children's books, though among these his best known were those on which he collaborated with Maurice Sendak, an illustrator whose innovative style rivaled that of Dr. Seuss. Sendak executed a series of delicate pen-and-ink drawings for Randall Jarrell's *Bat Poet*, published in 1963, that gained both of them wide exposure to an appreciative audience of children and, not surprisingly, many adults. Jarrell and Sendak had a final collaboration in 1964, working together on Jarrell's *Fly by Night*, his last children's book before his accidental death later that year. *Fly by Night*, published in 1976, is a classic bedtime story that comes as close to profundity as children's literature can. Jarrell's careful poetic prose, which has none of the condescending simplicity that is usually associated with children's literature, is perfectly matched to Sendak's imagery to produce this strange, wistful, and wise book.

Fly by Night is a story about a boy who can fly (well, actually something more like controlled floating), but only at night. The boy is naked, and there is an understated, infantile sexuality implicit in the idea of gliding naked through the summer night air. (Sendak borrowed the naked flight aspect of *Fly by Night* for his book *In the Night Kitchen*, which, although it was published before *Fly by Night*, had actually been written later.) Not only can the boy float wherever he wishes, but he also has the power to see dreams. On the night of the story he floats up out of his bed and into his parents' bedroom, where he can see their dreams "round and yellow" hovering just above their heads. He can see his dog's dreams as well. After floating through the house, he glides outside, where a full moon shines in the night sky. He sees mice dancing in the moonlight and a flock of slumbering sheep. "All of them except one are dreaming they're eating; that one is dreaming he's asleep."

He continues to glide through the night sky until he eventually floats to an owl's nest. Hovering near the nest he overhears the mother owl tell a bedtime story to her owlets, a bedtime story within a bedtime story. Then mother owl accompanies the boy back to his house, where he falls into bed and sleep. The following morning, when the boy wakes up, he remembers nothing of his adventures in the night, though when his mother hugs him the reader is reminded of the maternal love that the mother owl had for her owlets. The ambience and mood of *Fly by Night* leave us with a lingering sense of the charmed loneliness of existence: that even if we are ultimately alone, love and family sustain us.

Fly by Night introduced, possibly for the first time in children's literature, a romantic and sexualized relationship to night. It acknowledged the reality of children's bodies and the ambivalence most of them have toward night, but instead of sugar-coating this anxiety, *Fly by Night* celebrated it in an unexpectedly mature way. Sendak used this same formula for his next two children's books. He had become the darling of educated, psychologically aware parents in the sixties and seventies, parents who wanted the stories they were reading their children to have a developmental agenda beyond mere entertainment. His work was playful and just a little dark. What made Sendak's books innovative was the way he flirted with subconscious drives.

Sendak's biggest breakthrough, as both illustrator and writer, came with *Where the Wild Things Are*, also published in 1963. It was a runaway best-seller. Here his drawing style had evolved from the delicate illustrations of *The Bat Poet* and *Fly by Night* into the slightly rough-cast, vaguely Edward Gorey-esque approach that marked his later work. His color-washed ink illustrations with their crosshatched texture used a subdued palette of counterintuitive, earthy pastels that, at least for some of the more traveled parents who purchased his books for their children, must have been reminiscent of pre-Renaissance Italian painters.

Where the Wild Things Are is a story about a little boy, Max. It begins on a hot summer night just after Max (who is wearing a wolf suit replete with a wolf's head atop his own) has had an argument with his mother and has been sent to his room. Through the window of his bedroom we can see a deep blue summer night with a sultry, misty full moon rising. Somehow, as if his unconscious emotions project their own reality, Max's mood initiates a magic transformation—his bedroom changes into a lush, tropical forest. Eventually the walls disappear altogether and he begins to wander through a night jungle until he reaches a beach where a sailboat is moored, ready for him. He gets aboard and sails over the moonlit ocean, traveling for many days until he reaches a nocturnal island where the "wild things" live. The wild things are the embodiment of all the frightening monsters that lurk in the darkness, as well as the projected wildness of Max himself.

On the island of the wild things—which are elephant-size, plump, toothy, and look somewhat like large, animated gargoyles—Max meets dream creatures that more than match his willfulness. Here he is, still in his wolf outfit, pretending to be what these creatures very much are, but they are larger, and they outnumber him. Max is bold, though, and after his initial fears, he befriends them and soon he is riding the most fearsome of the wild things through the tropical night forest of this lost world. They have a wild night, and, before dawn, Max sails away, back to his room, where the full moon still hangs in the window. It is as if the visit to the island of the wild things was a dream that played out Max's anger and the wildness inside him.

The message of the book is clear: It's all right to be a little afraid as long as you are brave and allow your curiosity to move you forward rather than letting your fear get the best of you. Night can be enjoyed if it becomes an adventure.

This is certainly the theme in Sendak's later book *In the Night Kitchen*. Published in 1975, it was one of his most popular and famous titles. Gone are the crosshatches and muted colors. Here everything is bright, and his images are more iconic, though still very much Sendak. In the story a young boy, Mickey, is awakened in the middle of the night by a "racket." (We are never told what the "racket" is, though Freudians might hazard a guess.) Mickey gets out of bed and falls, or more nearly, floats downward in slow motion. He is naked and, perhaps for the first time in a children's book, his body is anatomically complete. The depiction of Mickey's penis was, and still is, too much for some. As a result, *In the Night Kitchen* has been removed from some library shelves and banned in a few schools. It wasn't Sendak's intention to test the limits of public morals, and he was genuinely surprised to hear that certain librarians had inked out Mickey's penis in their copies of *In the Night Kitchen*. Sendak was merely taking the exploration of childhood sexuality he had initiated with Jarrell in *Fly by Night* one logical step further. At any rate, Mickey drifts down through the starry sky past the moon, past his parents' bedroom, and into the night kitchen. The night kitchen is not a room, but is a hybrid of a kitchen and the city, with a deep blue, starry sky above it. Although the bakers look jolly and plump, resembling the dough they are kneading, the story takes a sinister turn when they grab Mickey and put him into the bread dough.

But Mickey isn't concerned; in fact, he likes it. I suppose the idea of being naked in warm bread dough is even appealing and titillating. But there is another theme, a sort of hybrid of Hansel and Gretel's close shave with the witch's oven crossed with the popular children's rhyme about bread dough ("Pat-a-cake, pat-a-cake, baker's man, bake me a cake quick as you can, pat it, and roll it, and mark it with B, and throw it in the oven for baby and me"), which has a tactile edge. Fortunately the bakers are more bumbling than sinister and Mickey jumps out of the dough. Here, as in *Where the Wild Things Are*, the protagonist overcomes anxious situations by taking control.

Mickey takes some of the dough and makes an airplane in which he flies to the Milky Way, which happens to be a two-story bottle of milk. Once there he leaps from the cockpit, naked, into the bottle of warm milk, where he swims undermilk, drinking at the same time, and saying at one point, "I'm in the milk and milk's in me." The book ends as the sun begins to rise and he returns home to his bedroom.

These four books are only my own favorites, and they are a very small selection from the hundreds of children's books about night from all over the world. (Amazon.com lists more than 1,585 children's books with *night* in the title!) From Japan, for example, comes the mysterious *Owl Lake*, by Keizaburo Tejima, the tale of a father owl who catches silver fish for his hungry children. It is illustrated with simple, old Japanese woodcuts. From France, *Night Creatures*, by Sylvaine Perols and Gallimard Jeunesse, is a beautifully illustrated survey of bats, owls, and other nocturnal creatures. Brazil's *How Night Came to Be*, by Janet Palazzo-Craig and Felipe Davalos, recounts a mythological tale of the first night, and from South Africa, the lovely *A South African Night*, by Rachel Isadora, describes African night animals that are active while children sleep. The Chinese folktale of the *Moonlady*, retold by Amy Tan, is an account of a grandmother who remembers when, as a young girl, she asked the Moon Lady to grant her a wish. This is perhaps an age of enlightenment for children's literature. Children today have an amazing array of extraordinary books that gives them wings to fly through the night, in their imaginations and their dreams.

Night Games

As children grow older, they begin to venture into the darkness unaccompanied to experience the night firsthand. I remember late spring nights, when my friends and I would meet after supper to play hide-and-seek in the twilight. As night deepened, the game grew more exhilarating. The seeker became a hunter, a night-blind carnivore, while the hidden players were like herbivores, trying to blend into the darkness. Everything about hide-and-seek conspired to build us into a peak of excitement: the anticipation of being caught, the triumph of calling "home free" when the kid who was "it" had no clue where you were hiding, narrowly eluding being tagged and the competitive desperation of a close race to "home," not to mention the delightful terror of being stalked while hiding in the night, and the secret triumph when the seeker walked within arm's length of your hiding spot without seeing you.

Sometimes I would use the darkness to my advantage and hide in outrageously exposed locations, very close to "home," aware that the night would hide me until whoever was "it" was farther away from home than I was. As we got older, a new element, a delicious thread of innocent sexuality, was woven into our hide-and-seek games. One of my first kisses was with the girl next door as we hid under a forsythia bush together on a warm May night. But any game we played at night—softball under the streetlights, "greenlight, red light" in a friend's driveway, or statue in someone else's backyard—had a special, thrilling undertone of illicitness. I like to believe that it was on those nights, when in the company of friends, we first began to feel the beginning of real adult freedom, a liberty initiated by the night.

5

TRIP THE NIGHT FANTASTIC: THE CITY AT NIGHT—9 P.M.

Cities, like cats, will reveal themselves at night.
RUPERT BROOKE

EVERY NIGHT HAS its own unique identity, and by nine o'clock the character of this particular night is well defined. Above the eastern skyline a surprisingly large full moon—magnified by the atmosphere—is rising redly, like a science-fiction planet. From the airport comes the occasional, distant rumble of jets taking off. Indoors children lay dreaming in their beds. Most adults are still awake and, although they cannot feel it, melatonin is beginning to be secreted by their pineal glands, preparing their bodies for sleep. At the same time, their basal temperature is also rising. Restaurant reservations are unnecessary now—the dinner rush ended half an hour ago and already the crowded tables are starting to thin out. Silence is beginning to reestablish its presence.

Urban wildlife is active in the residential sections of the city and house cats are outdoors and alert, or sitting in windows wishing they were outside. For them the night is filled with excitement: females in heat, territorial battles, and innumerable small, delicious rodents scurrying through gardens and lawns. Thousands of miles to the south, agouti, long-nosed southern relatives of the North American raccoon, are nosing through trash bins in the alleyways of Manaus, Brazil. Soon it will be curfew time in Trenchtown, a poor area in Kingston, Jamaica. TVs are a luxury here, and many of these tin-roofed shelters have no electricity. The sound of reggae drifts like hope through the tropical night air. Across the Atlantic, night is more established. Long-eared bats, which take prey from the earth as well as the sky, flutter through urban forests in London, England, occasionally dipping to the ground to snatch up a cricket. On the outskirts of Bombay, a king cobra slides into a kitchen through an open back door, its skin gleaming darkly against the tiled floor. It is a welcome guest in this Hindu home and is considered a good omen, though the residents will wisely give it a wide berth. Much farther southwest, in Amboseli National Park, Kenya, the same night is noisy with the grunts of lions, the whoops and cackles of hyenas, and the chirping of tropical crickets.

In the country during growing season almost everyone is indoors at nine o'clock, and the fields and woodlots are darkened. Farmhouses are islands of light riding the dark fields, their orange windows like domestic beacons in the twilight. A Portuguese fishing trawler is also a single point of light in the immensity of the North Atlantic Ocean. Onboard, the arc lights are blazing as the day's catch is packed with ice and secured in the hold. The work leaves no time to marvel at the unusually calm ocean, spread like a pink mirror under the rising moon.

Bright Nights, Big City

A city turns inside out at night. After the outflow of rush hour there is an hour's pause before the inflow of recreational pilgrims streams back into the city's core to fill the vacuum. The offices and buildings that teemed during day stand empty while the theaters, bars, discos, casinos, restaurants, opera houses, arcades, and concert halls begin to open. But only the core entertainment districts are active at night. Otherwise, the city is quiescent, and the municipal business of street cleaning and emergency road and transit repairs takes place without the hindrance of traffic. Hospital emergency wards are largely un-affected by night, though darkness brings an increase in the number of assault victims, people wounded by gunshots, and, in the small hours of the morning, the sad casualties of drunken crashes. There is a price for night's revelries.

Downtown, as Petula Clark sang, is where the lights are bright and you can forget all your troubles. And that's exactly what thousands of revelers are searching for, an antidote to their dreary occupations, something they can take back from life to make up for all the drudgery of their days. So they party hard, but their goal isn't just a good time. Night is also when they can transform into their ideal identities. The night city is like a fashion runway, a stage on which people can road-test their new selves—when the shy receptionist turns into a disco diva and the clean-cut dentist with a small office in a suburban strip mall becomes a Harley Davidson biker, kick-starting his "chopper" in a black leather jacket with steel studs. Night is when a retired pensioner in Dublin becomes a loquacious ranconteur in his neighborhood pub, and an office clerk in Taiwan transforms into a torch singer in a karaoke bar.

For those seeking a new life, a new partner, or just a one-night stand, the nocturnal crowds of the city become a vast array of possibilities, an infinite number of potential connections, exciting new friends, and star-crossed lovers. At 9 P.M. the night is indeed young: filled with hope, bravura, confidence, and a delicious anticipation of the unknown gratifications yet to come. In the entertainment district some nights of the week are quieter than others, though in a big town every evening has its ultimate destination—Wednesday night at the Paradiso, Thursday night at the Plaza, Friday night at the Rosewater Club. But Saturday night has always been the ultimate evening downtown. For those who work from Monday to Friday, Saturday night is a weekly vacation, an oasis in the middle of the weekend, insulated against the workweek by Friday on one side and Sunday on the other.

From Tokyo to Stockholm, Saturday-night fever is the virulent fuel of an urban, nocturnal carnival, and the singles mating game reaches a frenzied peak between 11 P.M. and 2 A.M. By 11 P.M. the bars, discos, and dance clubs are seismic with the subwoofers of powerful sound systems. Prospective dates bellow conversations at each other through the smoky air. Most available single women are surrounded by one, two, or sometimes three men. The competition is fierce—only the losers will go to bed alone tonight. Meanwhile, in the gay bars it seems everyone is dancing, and the air is hot and humid with sweat, cologne, and alcohol.

Nightlife is as old as civilization, or at least as old as torchlight. Babylonian palaces were lit up at night by torches, and Roman legionnaires stumbled drunkenly to brothels carrying portable oil lamps. In Roman Pompeii there were many all-night inns and bars, called *pervigiles popinae*, lit by large oil lamps. The fantastic nocturnal balls of the Medici court were brilliantly lit by candles and glass-hooded oil lamps. Kerosene lamps provided light for American saloons in the Wild West, but it wasn't until the advent of electric light that night was truly liberated from darkness.

Nightclubbing

The roots of the modern nightclub go back more than a hundred years, to the Parisian cabarets and vaudeville entertainments of the late nineteenth century. Originally *cabaret* was a French term for any business that sold liquor, but it began to take on a different meaning in the Montmartre district of Paris, when artists ventured out evenings in search of food, companionship, and conversation. One of the original Montmartre cabarets, the now famous Le Chat Noir (the black cat), began to put on amateur shows staged by some of their customers, who themselves were often poets, writers, and musicians—

the composers Eric Satie and Claude Debussy performed some of their earliest works there.

The timing of Le Chat Noir, which opened in 1881, couldn't have been better. In less than five years Paris was installing its first electric lights and both the Moulin Rouge and Le Chat Noir were electrified within the decade. Both clubs expanded to accomodate the deluge of new customers. Other clubs opened to cash in on the explosion of nightlife and soon Paris boasted dozens of cabarets and vaudeville theaters. The new technology of electric lighting also enhanced the artistry of cabaret performances as stage lighting and spotlights all became brighter, more colorful, and portable. The cabaret craze spread to other French cities and by the turn of the century to Germany.

It was in these German cabarets that some of the greatest song-writers and performers of recent history got their start—Marlene Dietrich and Kurt Weill both began their careers in Berlin cabarets. In the period between the First and Second World Wars these clubs became famous for their worldly hedonism, as well as their progressive acceptance of sexual deviation. Bob Fosse's film *Cabaret*, starring Liza Minnelli, was a fairly true-to-life rendition of pre-Nazi Berlin cabaret life, though purists insist that *The Blue Angel*, starring Marlene Dietrich, provided a much more accurate portrayal of the quintes-sential cabaret. Her smoky torch songs and androgynous sexuality romanticized the deep, urban night of the Weimar Republic.

While cabarets flourished in Europe, they had more difficulty taking root in America. Expensive New York City restaurants like Delmonico's and the Palais Royale had elaborate dinner shows that were called "cabarets," but these venues were not the real thing, and, at any rate, they were closed down in 1918 with the advent of Prohibition. Although catastrophic for the older, established dinner-show restaurants, Prohibition probably did more to accelerate the development of true nightclubs in America than anything else. It was

Prohibition that engendered the advent of the speakeasy. These illegal, secret bars were referred to as clubs that ostensibly catered only to card-carrying members. In reality, anyone with the money could get in if they could convince the doorman they weren't federal agents. The atmosphere of speakeasies was much more reminiscent of European cabarets, and they featured intimate performances steeped in the allure of illegal and forbidden pleasures. Some of the more organized speakeasies had house bands and regular entertainers. Singers such as Helen Morgan, who later starred in *Showboat*, got their start in speakeasies.

Vaudeville continued to be the main venue for variety shows through the Roaring Twenties but the growing popularity of movies eventually displaced it as America's favorite destination for an evening out. Vaudeville finally ended in 1932 at the beginning of the Great Depression, when its flagship venue, the Palace Theatre, stopped producing live shows and became a movie theater. Coincidentally, Prohibition ended a year later, in 1933, and speakeasies were suddenly legal. All they had to do was purchase a liquor license to become legitimate, and profitable, businesses. All these factors combined in the early part of the "dirty thirties" to set the stage for the rise of the chic supper clubs. These new clubs were on a scale previously unseen, even glitzier than the Palais Royale—they were indoor movie sets, with exotic tropical night themes and art deco interiors that reflected the glamour of Hollywood. Venues like the famous Copacabana and The Cotillion Room were fantasy destinations that provided lavish floor shows and imported some of the former stars of vaudeville, including Jimmy Durante and Sophie Tucker.

But all good things must end, and the upheaval created by the Second World War brought the supper club era to a finish. The American nightclub had to remake itself once again. The French and German cabarets had also met their demise with the war and, in the rubble of the postwar years, Europe looked to the United States for new models. After the war American supper clubs became smaller and more intimate—singers barely had a stage and they played smaller rooms that were often only large enough to shoehorn a dozen tables into them. This was partly a reaction against the grand scale of the prewar clubs and partly due to the economy of space in big towns where the new nightclubs flourished. In the 1950s these clubs became the spawning grounds for rising stars, with Broadway and TV scouts regularly scouring nightclubs for new acts. It was in the supper clubs and cocktail lounges of large American cities that the modern stand-up comedian was born—Lenny Bruce, Bob Newhart, and Woody Allen all got their start in nightclubs. Nightclubs were also where the countercultures of the late twentieth century began, the much lampooned beatniks of the fifties that in turn gave rise to the folk movement of the sixties.

At the same time, with the ascendance and then domination of world pop culture by rock 'n' roll, another, new kind of club began to emerge alongside the urban nightclubs. Dinner and dancing had became passé sometime in the late fifties. New environments were necessary and new nightclubs, like the Peppermint Lounge in New York City (made famous by the "twist" dance craze), sprang up. Rock 'n' roll migrated to Europe, and Paris, which earlier had spawned the cabarets, became an innovator again, with the inception of the discothèque, later shortened to *disco*, which invaded the world with a vengeance in the mid seventies, creating a whole new genre of rhythm and blues to go with it.

In the entertainment economy of the new millennium, the nightclub has undergone yet another transformation, only this time it has recovered all its previous incarnations. Films and TV, by romanticizing and resurrecting the past, have contributed to a retro culture where what once was is once again. Here in the new millennium, nightclubs are still going strong, though they compete with discos, live-music bars, hip-hop lounges, cocktail lounges, raves, DJ halls, karaoke bars, dance clubs, and singles bars. The entertainment district of any large city, anywhere in the world, has available all the various stages of nightclub development, at least as retro references in nostalgic or theme environments like Harry's Bar in Venice or the resurrected Palais Royale in Toronto. These nocturnal playgrounds are the outposts of a burgeoning colonization of darkness, one in which night becomes the sheltering backdrop for our brightly lit, urban amusement parks. The curtain of darkness, at least in the cities, is being drawn back.

The Dark Frontier

And the best of all ways
To lengthen our days
Is to steal a few hours from the night, my dear.
THOMAS MOORE

"Cities of the red night," William Burroughs called them, the world's largest cities: London, Rome, Buenos Aires, New York, Mexico City, Calcutta, Paris, Tokyo, and Shanghai. Cities where, on cloudy nights, the light from innumerable streetlights, neon signs, office buildings, and homes turns the sky deep red, though this might be an over-statement. The nightglow of the city on low clouds is more like a subdued electric pink than red; in fact, depending on the weather, on some overcast nights the urban glow is a bruised mauve—other nights it is a fleshy, neon salmon. This reflected light, absorbed deeply into the clouds, rivals moonlight for nocturnal illumination, though it is diffuse and casts no shadows. A dark room illuminated solely by city glow through a large window is bright enough to see most of the objects in it. Sometimes there can be city glow even on clear nights if there is enough humidity.

A few years ago I drove back to Toronto from a friend's cottage on an overcast summer night. The traffic was light and I was in a driving trance, letting my thoughts drift the way they do when you are driving alone at night. I had turned off the radio and navigated in relative silence with only the sound of the wind sliding by the windows (and the illumination of the instrument panel) to accompany me. When I was a little less than fifty miles away from the city, I began to see a reddish glow just above the vanishing point of the highway. It was mesmerizing. I kept glancing up to look at the sky and by the time I was twenty-five miles closer, the immense, pink glow of the city stretched from the western horizon to the eastern horizon. Even though I knew that the glow was the aerial reflection of the lights of my city, it seemed, on a purely animal or visceral level, almost alarming, like a conflagration or a giant inferno. It was one of those moments where you stand outside your own time, your own culture, and see things for what they are. Was anyone else traveling south on the highway as excited by the spectacle of light that was about to consume him or her? Probably not. People seldom look up in awe at the city sky, and generations have been born since Edison first lit up New York. Our wealth of light is rarely noticed, much less appreciated.

Nonetheless, the electric blaze of the city at night is one of the most fantastic sights of our times. It is all the more impressive if you consider that little more than a century ago this extravaganza of light was impossible. Yet at the same time, our urban pageant of light appears so permanent, so monumental, it seems impossible that it hasn't always been so. After all, the glimmering skyscrapers that make up the central spectacle of any big city can be seen from more than thirty miles away! Viewed from an elevated vantage even closer, say the outdoor terrace of a rooftop patio two miles from the city center, the office buildings that form the central peaks of the city spire into the dark sky like angular fountains of light. As the night progresses, the luminescent columns seem to dissolve into irregular vertical chessboards, mosaics of light and dark. Some floors become completely black and disappear while others remain lit, until the towers look as if they're composed of floating layers of fluorescent strata. The profile of the city slopes rapidly down from the financial section, through the ring of shorter buildings that surrounds the core megaliths, and spreads out from downtown into a sea of sparkling streetlights and plazas. This incandescent plain is interrupted with sporadic atolls—clusters of apartment buildings or office towers that punctuate the grid of glimmering streets like icebergs of light on a glowing ocean.

In his 1987 book, *Night as Frontier: Colonizing the World After Dark*, Murray Melbin claimed that darkness was, and still is, a hostile realm that we are conquering with technology. Just as we invented boats to cross oceans, fire and central heating to colonize the cold climates, airplanes to conquer the air, and Aqua-Lungs to penetrate the oceans, so we invented fire and electric light to conquer night. Unlike other realms, though, night is not just a physical, three-dimensional territory. Melbin claims that the colonization of night is the beginning of a new expansion for us, one entirely unlike any previous territory we have entered. For the first time we are expanding into the fourth dimension, the dimension of time.

In a sense we are like miners, tunneling with light into the bedrock of darkness. Artificial lights carve tunnels and caverns out of the night, spaces in which we can operate as if it were day. In another sense, reclaiming light from the darkness is similar to reclaiming land from the ocean. The city at night is like a *polder* land, the Dutch term for fertile farmland taken from the sea by pumping seawater out of diked fields. In these luminous polders we can continue our daytime pursuits and reclaim new time from the night. According to Canadian media theorist Marshall McLuhan: "The electric light ended the regime of night and day, of indoors and out-of-doors . . . Cars can travel all night, ball players can play all night, and windows can be left out of buildings." In another way, though, we are confined by light. We cannot wander outside of it, unless equipped with night vision. In a sense we are like light termites who require light the way real termites require humidity. On the face of it, the termites' survival strategy seems doomed because, for all their global ubiquity and their fearsome reputation for devaluing real estate, termites are fragile, exotic creatures. They can only exist within a very limited range of high humidity—anything outside of that range will cause them to dry out and die in a matter of minutes, like astronauts without spacesuits. That is why termites have to build airtight, sealed corridors anywhere they want to go. As an engineering project it seems counterproductive. If a single termite had to build a tunnel to any particular food source, it would starve of malnutrition and exhaustion before it even reached it. But, as part of a larger colony, with allocated tasks, a termite coordinates its behavior with that of its sisters.

Similarly we require a larger colony to build the infrastructure, streetlights, and roads for individuals to use. We are prisoners of artificial light who cannot stray outside its perimeter unless we can take our "light tunnel" with us in the form of night vision or flashlights or the headlights of our cars. Perhaps that is why, when we're on nighttime automobile trips between cities, we feel such a sense of release and freedom when the last lights of the city expressways are left behind and we're immersed in the cozy glow of our instrument panels and headlights within the larger darkness of night.

The End of Darkness

The discovery of fire, at some time deep in our prehistory, was the beginning of our conquest of night. With fire we were no longer completely blind in the darkness, and when the first Paleolithic human picked up a burning branch and walked with it, perhaps to frighten off a hungry animal or simply to see something beyond the ring of light from his fire, we began to penetrate the night. For the first time we were no longer bound by the cycle of light and dark. We started to become creatures of the night.

From a burning firebrand it was a short evolution to the torch. According to archaeological evidence, torches have been around for at least eighteen thousand years, and were certainly in use around the time the Cro-Magnons were painting their delicate and sophisticated likenesses of animals in the Lascaux caves in southern France. The ceilings of these caves are still blackened in many places with the soot from their torches. It probably wasn't long before another ancient inventor noticed that the grease from animal fat burned longer than wood, and the jump from the fat-soaked torch to the portable oil lamp with a wick and a bowl reservoir probably took place fairly quickly. But it was still a long time before many lamps came to be gathered in one place, and it wasn't until the advent of the first cities that the darkness of night was first seriously challenged.

If travelers from another planet had arrived at earth's night side five thousand years ago they would have seen no lights, no evidence of intelligent life, at least from their vantage just beyond our atmosphere. On the day side it would be obvious that life had taken hold on earth's surface, forests and grasslands would be immediately visible, but our space travelers would have to search the entire planet with a powerful telescope to see the first large-scale structures that signaled the birth of civilization in the Near East. A thousand years later the night side of the planet earth would reveal a different story. The first great cities of Mesopotamia would have already begun to glow in the darkness and our mythic travelers would, for the first time, see pinpoints of light on the dark side of the earth that were not forest fires or volcanoes.

The plentiful availability of vegetable oil, a consequence of large-scale, organized agriculture, revolutionized oil lamps around five thousand years ago. No longer dependent on animal oils for fuel, portable, earthenware oil lamps became cheap and efficient. They have been found in Sumerian royal tombs that are more than forty-six hundred years old and they were probably in use in the first great cities. The windows of temples and certain municipal structures would certainly have been lit at night, not to mention hundreds of homes of wealthy citizens. By the time of the Roman Empire, less than two thousand years later, when oil lamps were mass-produced and the candle had been invented, even the lowliest apartment blocks would have had interior lighting. For our interplanetary travelers, Rome of 40 B.C. would have been a large glowing point of light in the night side darkness, even though it had little street lighting.

Throughout the next fourteen hundred years, after the decline of the Roman Empire and the rise of the Dark Ages, our mythic travelers from space would have seen dim night lights in some of the larger cities in Europe, Asia, and, eventually, the European colonies, but these would barely have matched Rome's lights at their brightest. Meanwhile the oil lamp was refined into the glass-shrouded kerosene lamp, and as the Dark Ages gave way to the Renaissance, social reform in Europe began to affect the way that cities were governed. It was with the beginning of enlightened urban policy that ancient Rome's nocturnal brightness was finally surpassed in 1667, when the king of France, Louis XIV, decreed that the streets of Paris were to be illuminated by large candles enclosed within glass lanterns. A little more than ten years later Paris had sixty-five hundred lanterns burning at night, indeed earning it the name "city of light." From a vantage high above the atmosphere, Paris would now shine more brightly than any city had before. The final ramparts of darkness were being surmounted and consequently more and more city dwellers were beginning to venture out into the darkness to participate in the beginning of a great urban nightlife.

In 1776 London followed the example of Paris and installed five thousand oil lamps in the streets, and within two years this number had doubled. Now two, much brighter cities would glow on the dark side of the planet. The advent of street lighting marked the beginning of the most aggressive phase of the colonization of night. In the beginning, street lighting was a novelty, with little impact on the waking and sleeping patterns of all but a handful of people, but eventually, as factory owners and businessmen realized that night-shift work could increase their profitability, more and more people began to work and live at night.

In the century that followed, there was an unrivaled acceleration in the progress of science and technology. Between 1807 and 1826, London, Baltimore, Paris, and Berlin all installed gas lighting in their streets. Now a new feature appeared on the night side of the planet, a luminescent geometry to delight our hypothetical space onlookers and one that would be familiar to anyone who has flown over a city at night—the gas-lit cities began to develop into glowing grids of light. Bright street lighting reinforced the linearity of streets, and cities began to develop into crystalline, luminescent lattices—at least when seen from above. Urban nightlife was also expanding, as stores, restaurants, and bars were staying open later and later to cater to the nighttime population. But there was still a long way to go before cities could rival the electric displays of the twentieth century.

By today's standards the city streets of the nineteenth century were still relatively dark. Charles Dickens, a tireless explorer of the city of London at night, chronicled the dim light of the gas-lit streets, describing it as crepuscular and insufficient. But even Dickens had no idea of the change that was going to occur before the century had finished. In the last half of the nineteenth century the pace of invention became delirious. Dickens's era was the time of the great inventors: Nikola Tesla, Alexander Graham Bell, Samuel Morse, all invented devices that changed the world. But, in terms of their impact on night, none rivaled Thomas Edison, who in 1878 constructed the first electric lightbulb. Four years later, in 1882, he installed the first public electric-lighting systems in New York and London. The radical effect of the electric lightbulb cannot be overstated. More than any lighting technology that preceded it—candles, oil lamps, or gas lighting—the electric light revolutionized the night. Now the darkness, at least in cities, was in full retreat.

Within ten years the great cities of Europe became brilliant citadels of light. People who witnessed the transformation were awestruck. One of the consequences of electric light was a huge demographic shift, as more and more rural villagers were literally mesmerized by the spectacular light shows of the great cities; electric light popularized the city to country dwellers. But artists were as enthralled by this transformation as anyone. James McNeill Whistler was lucky enough to witness the transfiguration of cities by electric light in the late nineteenth century. He wrote, "The warehouses are palaces in the night, and the whole city hangs in the heavens, and fairyland is before us." In 1909, F. T. Marinetti wrote about his experience of electric lights in the founding manifesto of futurism. By his time even automobiles had electric lights and he described double-decker buses as being "ablaze with colored lights, like villages on holiday," or the brilliantly lit nocturnal shipyards of Italy as "blazing with violent electric moons."

But this colonization of darkness also jump-started a new wave of urban nightlife. The futurists in their second manifesto, signed by Umberto Boccioni, Carlo Carrà, Luigi Russolo, Giacoma Balla, and Gino Severini, described this new recreational frontier of night in the great electric cities as "the exciting new psychology of night-life; the feverish figures of the bon viveur, the cocette, the apache and the absinthe drinker." In a later manifesto they claimed, "The suffering of a man is of the same interest to us as the suffering of an electric lamp, which, with spasmodic starts, shrieks out the most heartrending expressions of color."

The twentieth century saw the greatest advances in nightlife, as hospitals, police stations, factories, military operations, supermarkets, night courts, discos, nightclubs, theaters, and television and radio stations transformed Western, industrialized nations into all-night edifices. By the turn of the millennium, the phrase "24–7" was coined to describe a world that never slept, that was operating twenty-four hours a day, seven days a week. And our glittering architecture keeps growing. For pure, awe-inspiring grandeur, nothing beats flying at thirty thousand feet over New York City on a clear night. The sheer size of it, the glowing, jeweled crosshatching of streets and shimmering skyscrapers, the bright angular grid of streetlights stretching right to the horizon, even from such a great height.

But this light show is expensive. We use turbine generators to squeeze electricity out of coal, oil, natural gas, uranium, and gravity. The municipal street and highway lighting alone costs the city of Toronto, an average-size city of three million, $14,336,931 a year, which means that an average night costs City Hall a whopping $39,280! And that's just the power bill. There are many other nocturnal costs for the municipality, including security guards, skeleton staff, and police helicopters, and the hidden costs of light pollution, such as the health care expenses that arise from sleep disturbances and the disruption of normal circadian rhythms. In the Neolithic era night used to exact its price with nocturnal attacks of predators; now it takes a bite out of our wallets.

Light Pollution

Handicapped by night blindness, the majority of us live in illuminated urban oases, though the August 2003 blackout in North America gave people there a taste of what it was like less than two hundred years ago when nighttime activities were restricted to the small circle of light cast by candles. Over the past hundred years electric lights have proliferated across the planet until the night side of our planet twinkles with terrestrial constellations. Great networks of cities are linked by glowing threads of urban sprawl—they twinkle on the night side of the planet like a science-fiction fantasy of light. If our hypothetical space travelers had returned to earth at the turn of the millennium after a one-hundred-fifty-year absence, they would have been astonished at the transformation of the night side of the planet. Whole sections of continents, particularly North America and Europe, would be alight in the darkness.

But there is a price to pay for this "day in the night." There are people who have been raised in the cities who have spent their whole lives without seeing the transcendent majesty of a crystalline night sky. In a natural night sky about thirty-five hundred stars should be visible to the unaided eye, but in a city, even in a dark yard, only about fifty stars are visible. Some of the more memorable stories that came out of the great blackout of August 2003 were from city dwellers who, for the first time, saw the city skies filled with bright stars. Two-thirds of the world's population lives in cities—in Europe, North America, and in parts of Asia that number is closer to 90 percent. In the larger urban areas it never really gets dark at all. As anyone who lives in a city will tell you, night is more like a period of dimness than of pitch-blackness. When city dwellers turn off the lights in their bedrooms, after their first momentary night blindness, they gradually become aware of the background glow of the city night. If their curtains are not drawn, then their bedrooms are flooded with a pale violet, electric light.

Astronomer Pierantonio Cinzano of the University of Padua, a light-pollution activist, says that about 10 percent of the world's population have lost much of their night vision as a result of this urban glow. In North America that figure is more like 60 percent. But that's not all. Aside from the many animal species affected by light pollution—the thousands of migratory birds, disoriented and killed by skyscraper lights, the female sea turtles trying to lay their eggs on illuminated parking lots and highways, as well as the millions and millions of insects that are hypnotized and killed by lights—there is frightening new evidence that light pollution makes us sick.

Richard Stevens, an epidemiologist at the University of Connecticut, thinks that light at night disrupts our circadian hormone levels, and as a result we are becoming ill. In a study he coauthored in 2001 he wrote that there was "mounting evidence to suggest that disruption of the melatonin rhythm may lead to chronic fatigue, depression, reproductive anomalies, and perhaps even cancer." He also found that the risk of breast cancer was raised by 60 percent for women who worked night shifts and that, conversely, the rate was 50 percent lower for women who were blind. Stevens suspects that because melatonin inhibits estrogen from accelerating the growth of breast cancer cells, its depletion causes a greater susceptibility among women who don't get enough darkness. Stevens found that even women who didn't work on night shifts, but merely slept in a moderately lit bedroom, still had higher risk levels for breast cancer.

But does the same effect hold true for city dwellers who have mildly elevated light levels even though they sleep in relatively dark bedrooms? According to David Blask, a scientist at the Bassett Research Institute in Cooperstown, New York, it does. After seeing the remarkably positive effects of melatonin injections on cancer patients, he decided to conduct experiments with three groups of rats to determine the carcinogenic property of light pollution. One group had round-the-clock office lighting, another had office lighting by day and total darkness by night, while the third group had office lighting by day and a slight amount of nighttime light. He was surprised to see that the tumor growth in the rats exposed to the very low nocturnal light levels ("It would be the equivalent to being in a completely dark room, with a small light coming through a crack in the door," Blask said) was the same as in the ones exposed to continuous office light. Though humans are not as sensitive to light at night as rats, he says, "There is enough evidence right now to be cautious at any rate and certainly limit our exposure to bright light at night."

But the popular recognition of light pollution is lagging behind recent discoveries, which is frustrating for dark-sky activists. First and foremost among these is the International Dark Sky Movement, the Greenpeace of the night sky, which is trying to organize international legislation to curtail light pollution. They are not the first to rail against urban lighting. The first concerted political protest against street lighting was during the French Revolution in 1789, when roving Parisian mobs systematically smashed street lanterns, which they regarded as surveillance devices of the repressive regime of Louis XVI. Before the guillotine was invented, the revolutionary mob used to hang their victims from the street lanterns as a sign of their rejection of monarchial authoritarianism. Perhaps their fear of the relation between street lighting and a centralized, authoritarian state was correct; after all, the first urban police department was deployed in London in 1829, less than two decades after gas lighting had been installed on public streets. In fact, street lighting and policing have always had a symbiotic relationship with each other. After the introduction of electric lighting in New York City, the chief of police said, "Every electric light erected means a policeman removed." This, sadly, has hardly proved to be the case.

Because of their benefits to citizen safety, street lighting, electric signage, and bright office windows have also been allowed to proliferate unchecked. But recently the dark-sky activists' message has started to percolate into mainstream culture; an episode of *The Simpsons* was even dedicated to light pollution. But there is still a long way to go. The International Dark Sky Movement believes that light pollution should be getting the same attention as other environmental ills, particularly given the fact that over the past thirty years light pollution has been increasing by 10 percent a year.

Some communities are beginning to see the night, though. There are a handful of dark-sky preserves starting to be set up around the world, and among these, perhaps one of the most extraordinary is Vicuña, Chile. Vicuña is situated beneath the world's clearest night skies and it is adjacent to some of the world's foremost astronomical observatories, including the Cerro Tololo Inter-American Observatory, located on mountaintops just outside of the city. A number of years ago Vicuña's burgeoning streetlights began to soak into the atmosphere so deeply that they threatened the astronomical views of the nearby observatories. A group of worried astronomers took their concern to the town officials and tried to work out a solution.

With an eye to tourist dollars as well as a conciliatory offer from the astronomers to build a small, tourist-accessible observatory, the city agreed to install caps on their streetlights that focused light downward, thereby preventing glare from shining up into the night sky. The strategy worked and within a year the city managed to cut its light pollution in half. True to their promise, the astronomers donated a fine twelve-inch reflecting telescope to the city in return. The telescope, housed in what is now called the Mammalluca Observatory, has attracted more than thirty-five thousand tourists since it opened in 1998.

That's exactly what the International Dark-Sky Association has been campaigning for around the world. Dave Crawford, the executive director of the IDA, cites Tucson, Arizona, as another good example of a city that has instituted night-sky protection ordinances. Tucson installed light caps on all of its streetlights and enforced aggressive bylaws to ensure that private commercial signs and lights also conformed to its light-pollution standards. As a result, even though Tucson has a population of approximately eight hundred thousand, all of its citizens can see the Milky Way on a clear night.

Night Watch

Watchman, tell us of the night,
What its signs of promise are.
JOHN BOWRING

By 9 P.M. Ed Villamere, a constable with the York Regional Police of Toronto, has been at work for two hours of his twelve-hour shift. Ed is an affable, fit man in his early thirties. Born in Quebec, he is fluent in French and English and has a friendly smile, though occasionally you can see flashes of deeper steel in his eyes. Confidence and authority suit him. He works the night shift in a police department that spans five subdistricts and two city worlds; downtown and suburban. When he checked in at 6:30, his staff sergeant assigned him a vehicle and his patrol area for the night, a cross-section of the edge of the city measuring ninety-three square miles. He has no partner.

His nights are almost always busy, and for the first eight hours of his shift he usually goes from call to call without a break. Constable Villamere likes to listen to the routine banter of the dispatcher on his police band radio, as well as the voices of the other officers working his region, some male and some female. They are just voices in the night, but they make him feel less isolated and alone; it means that someone else is in exactly the same position as he is—half target, half hero. At 9 P.M. Ed is feeling fairly crisp and he won't need a second coffee for a few more hours.

Police work, Ed told me, consists of long hours of answering 911 calls, responding to domestic disputes and break-ins as well as dealing with traffic violations. Occasionally this routine is interrupted by brief bursts of high-voltage action. He went on to say that statistically the greatest number of break-ins, assaults, and car thefts take place under cover of darkness. As well, most highway accidents occur between midnight and 6 A.M., the period of greatest driver fatigue. Ed has seen his share of grisly sights in the dark: what was left of a man who jumped off a bridge, witnessing an official postmortem in a midnight coroner's office, a drug overdose victim staring with glassy eyes at infinity.

I asked Ed if he was ever in any police pursuits. Yes, he replied, there was one just last year. It started around 9 P.M. and several cruisers were involved. He said the decision to give chase rests with the staff sergeant, who has to weigh all the factors concerning public safety, which, Ed was careful to insist, is the number-one priority. In this case they got the go-ahead and the pursuit lasted well over six miles and ended with a radio-coordinated trap. Another cruiser laid a spike belt ahead of the fleeing car and it blew out two of the suspect's tires. Ed and another patrol car caught up to the suspect and surrounded the car, guns drawn. For some reason the driver wouldn't get out of his vehicle, Ed told me with a note of incredulity in his voice, and then, to make matters worse, the man began rooting around in his car, possibly for a weapon. Remaining calm, Ed and the other officers didn't open fire. Instead they broke a window and filled the car with pepper spray, eventually forcing the driver out. It turns out he had been trying to hide his drug stash.

The only thing Ed doesn't like about working the night shift is that he is on the opposite schedule to the rest of the world. As a result, he exists in a lonelier city with a much smaller population. If that weren't bad enough, he has to sacrifice his own sleep to do errands during the day that most people take for granted. Demographically, night-shift work is increasing every decade, but night-shift workers are still an overlooked minority in a society that caters to day jobs. According to statistics released by the U.S. Department of Labor in 1997, 3.5 percent of all full-time and salaried employees work nights. If you include part-time and seasonal workers, the percentage jumps to 17.4. As well, people who trade night for day also change their physiological rhythms. Security guards who patrol empty buildings or corporate property at night, night nurses, cab and ambulance drivers, and police officers all make sacrifices, sometimes with major repercussions. A Finnish study revealed a higher rate of heart attacks among night-shift workers. They also drift out of friendships, their marriages fail more often, and they lose touch with the daytime world. American figures show that the divorce rate in marriages with one or both partners working night shifts is 7 to 10 percent higher than the national average. Because night-shift workers fight their own circadian rhythms, they sometimes have problems with insomnia, and their immune systems can also be negatively affected. The operator of the *Exxon Valdez,* the oil tanker that ran aground on the Pacific Coast of North America, was suffering from a sleep disorder, and it is known that one of the causes of the Chernobyl meltdown was the sleep disorder of a highly placed technician who was on the late shift. In fact, every nuclear accident in the world, including the one at Three Mile Island, occurred during the night shift.

There are other, less obvious though equally insidious consequences of night shifts. In large financial firms and the corporate headquarters of global companies, night managers are often delegated an authority they wouldn't hold during the day when more senior executives are present. This provisional delegation means that emergency decisions might be made by individuals not completely qualified to make them, and sometimes even routine corporate decisions that night managers make can be inconsistent with the daytime policy of the same company. Frighteningly enough, this same inconsistency also visits crucial government and military installations where fast decisions on matters of tremendous national security have to be made. Tired military personnel have their fingers on the nuclear triggers deep in NORAD's hardened, mountain bunkers. The presidential override codes await only a verified, dual-source confirmation of the president's incapacity.

Ed's job, fortunately for him, doesn't seem to have caused him any significant physical symptoms. He gets a good day's sleep and he doesn't have any residual alertness problems with deep-night shifts. If he does get a little drowsy, he stops off at an all-night doughnut shop. What about doughnuts? I asked him. Is it true that police eat doughnuts all night? That's just a myth, he replied, "I rarely eat doughnuts." The reason you see police cars at night at doughnut shops, he went on, is because those are the only places they can fuel up on caffeine: "Sometimes, if I really need it, I'll drink tea and Coca-Cola as well as coffee. I really enjoy this job," he says, "so much that I tell people that it sure beats working." Ed smiles as he delivers this homespun witticism, but as a private citizen I am reassured to know that police like him are watching over us in the night, that he is indeed the watchman of the night.

Ladies of the Night

Wild Nights—Wild Nights!
Were I with thee
Wild Nights should be
Our luxury!
EMILY DICKINSON

If it is not the oldest profession, then streetwalking is certainly an ancient one. Ladies of the night have plied their profession for thousands of years, and in the long and often hidden history of the sex trade, concubines, prostitutes, and courtesans have reflected the moral climates of their times. There have been periods of great tolerance, such as in ancient Greece, where courtesans, or *Hetairai*, as they were known, could gain noble status, and in Rome at the time of the Caesars, when prostitution was accepted and the government levied taxes on prostitutes. But there have also been periods of proscription and intolerance, even recently, such as Mexico's prohibition of 1930, when prostitution was aggressively outlawed for many years. But the sex trade has always survived regimes of repression and has, over its long history, become a sort of de facto guild, with as many different levels of income as a modern corporation. The men and women who ply their trade in the streets serve a very different clientele than those who publicly escort the very wealthy. Indeed, even with contemporary elite escorts, such as those that Hollywood procuress Heidi Fleiss organized, the delay between payment and service, as well as the length of a contractual arrangement, may blur the borders between purchased sexual favors and a practical, domestic relationship, albeit somewhat irregular. In the past there have been many instances in which courtesans, the escorts of nobility and aristocrats, have transcended their financial arrangements and become the life partners of their clients.

The tradition of the courtesan began in the great Mediterranean civilizations of Mesopotamia and Egypt. Ritual religious concubines were an intrinsic part of several Egyptian rituals, and sacred prostitutes were employed in some temples in ancient Greece. By the fourth century B.C. the interrelationship of wife, mistress, and civil concubine (a woman who often stayed in the family home and shared domestic duties as well as a carnal relationship with the head of the household) was an important aspect of Hellenic society. Athenian orator Apollodoros described this system very succinctly: "We have courtesans for pleasure, and concubines for the daily service of our bodies, but wives to produce legitimate offspring and as reliable guardians of our household property." This arrangement continued uninterrupted into Roman society, though with the advent of Christianity in the third century A.D., it changed drastically. Because Christian canonic law forbade married men from having concubines, the quadralinear relationship of husband, wife, concubine, and courtesan became a triangular one, with only the courtesan, husband, and wife. Henceforth the courtesan had to play the role of both regular concubine and occasional mistress.

For nonpatricians there were always common prostitutes, and Christianity, far from spelling the end of prostitution, merely changed the rules. During the Byzantine period of the Holy Roman Empire, prostitutes continued to work throughout the Mediterranean region, particularly in the Holy Land. There were even prostitutes that specialized in enticing monks, targeting those orders that lived in remote caves in an attempt to escape the temptations of the flesh. (Of course they were going crazy from celibacy.) Some Christian sects were more accepting of the role of lust and sexuality. Part of the Nag-Hammadi Codex, a collection of Gnostic verse from Middle Egypt of the fifth century A.D., was written from the viewpoint of a harlot who proclaims that:

> *I am She whom you honour and disdain.*
> *I am the Saint and the prostitute.*
> *I am the virgin and the wife.*
> *I am knowledge and I am ignorance.*
> *I am strength and I am fear.*
> *I am Godless and I am the Glory of God.*

This is perhaps the first recorded version of the whore/madonna complex that seems to fascinate so many men, and the central persona that the pop singer Madonna plays on so adroitly.

Around the same time as this Gnostic verse was written, one of the first great courtesans of history was making the night her personal empire and a road to divine glory and riches. The legendary Theodora, consort of the Byzantine emperor Justinian, was a courtesan who came from a common brothel. According to Roman historian Procopius, Theodora first started work as a child prostitute in a brothel in

Constantinople. Before she reached puberty, she disguised her gender in order to become a catamite. After she reached sexual maturity, she first became a low-level courtesan and then graduated to performing as an "actress," stripping in front of audiences and performing sexual acts onstage. A favorite at aristocratic banquets, Theodora gained notoriety for offering herself to all in attendance, including slaves. After a brief love affair with a young noble who had been appointed the governor of Pentapolis, Theodora began mixing with the rich and powerful. When she moved to Constantinople from Alexandria, she met Justinian, an heir to the throne of the Eastern Roman Empire, who also had common roots. Justinian fell deeply in love with Theodora and bestowed patrician status upon her. Then, after the death of the empress, Justinian forced his uncle, Emperor Justin II, to repeal the law that prevented courtesans from marrying senators. Justinian then married Theodora, who was thereby elevated to the ranks of nobility. In short order Justinian became co-emperor with his uncle and then, when his uncle died, he became sole emperor and Theodora became his queen. Theodora was a benevolent empress who championed the rights of women in matters of divorce and property rights. She stopped the practice of infanticide and the death penalty for female adultery. Remembering her past, she closed many brothels and created convents for the dispossessed prostitutes, hoping to better their lives.

It wasn't until the Renaissance that any courtesans rivaled the political success of Theodora. Although none were as powerful as Theodora, several influenced the policies of their lovers and patrons. In fact, the courtesans of the Renaissance not only challenged the societal standards of their time, but also the politics of gender. Women like Anne "Ninon" de L'Enclos (1620–1705) of France were protofeminists, whose independence and quest for equality often got them into trouble with the patriarchal societies of their day. In other ways the history of the European courtesans parallels the geisha tradition in Japan, where women were paid to be articulate, intelligent, and well-educated nocturnal escorts, as well as uninhibited bedroom partners. Like their geisha counterparts, the female courtesans of the Renaissance were often as well educated as men and traveled in aristocratic circles. Not only were they allowed to snub social tradition by sharing the prerogatives of men—education and access to libraries—but some of them even published their own writing, an unheard of practice at the time. Among these proficient, musically trained odalisques was Tullia d'Aragona (1508–56), an Italian courtesan and poet from Rome who published a book of poems called *Rime* in 1547. Cleverly, this book was dedicated to Eleonora, the wife of Cosimo de' Medici, perhaps as a ruse to make sure Eleonora never resented that it was Cosimo with whom Tullia was involved in her capacity as courtesan.

Another Italian poet and courtesan was the famous Venetian Veronica Franco (1546–91), whose life was made into the 1998 movie *Dangerous Beauty*. Franco, herself the daughter of a courtesan, was raised as a normal child and married at an early age. Disastrously, her marriage ended badly and she had to avail herself of her mother's profession, a vocation at which she excelled, soon rising through the ranks of courtesans to become a lover of King Henri III of France among other notable aristocrats. Like Tullia d'Aragona, Veronica Franco was also a poet, and wrote two volumes of verse: *Terze rime* and *Lettere familiari a diversi* in 1575 and 1580, respectively. A wealthy philanthropist, she founded a charity for courtesans and their children, though she was forced to leave her home in 1575 when the Black Death swept through Venice. When she returned, not only had her house been gutted but she was accused of witchcraft by the Inquisition, a charge of which she was eventually acquitted.

Many of the elite ladies of the night were, as Veronica Franco had been, courtesans in major European courts. Ninon de L'Enclos was perhaps the most famous of these. Born in Baroque Paris, she was nicknamed Ninon by her mother. When she was a little girl, Ninon's father was exiled and her mother died when Ninon was in her early twenties. She then entered a convent but stayed only a year because she had come under the intellectual influence of epicurianism and Montaigne, teachings that maintained that the sensual pleasures of both the intellect and the body were supreme. In Paris she became the darling of the salon scene, influencing the young Molière among others. The night became Ninon's grand theater of desire, and the Baroque courts were her domain. Ninon was also an author and a social philosopher, and in books like *La coquette vengée* (the flirt avenged) she described how a refined existence could be led in the absence of religion. Her wit was often quoted, and at one time she quipped, "Much more genius is needed to make love than to command armies," and "We should take care to lay in a stock of provisions, but not of pleasures: these should be gathered everyday."

As a courtesan Ninon was somewhat of a dilettante, and she only entered into financial arrangements with her suitors on a few occasions. This would not have damned her in and of itself, since courtesans were an integral part of court life, but her high profile and her outspoken views on religion caught unfavorable attention, and eventually, at the behest of the queen, she was imprisoned in a convent in 1656. Fortunately, the former queen of Sweden, Christina, championed Ninon's cause and eventually secured her release. Ninon retired as a courtesan in the late 1660s and devoted herself entirely to her evening literary salons, befriending the playwright Jean-Baptiste Racine among others.

In the next century the French produced another great courtesan, Madame de Pompadour (1721–64), formerly Jeanne-Antoinette Poisson. Like Ninon before her, Madame de Pompadour grew up without a father, who had likewise been exiled when she was young. She was well educated and refined, and when she came of age, she married an aristocrat. Her marriage gave her entrée to the rarified atmosphere of the royal court and she quickly realized that to climb in this elite society she had to become a de facto courtesan, though not just anyone's. She had her sights on the king himself. She was a legendary beauty and a marvelous dancer, and she became the jewel of the evening balls. It was at one of these nocturnal soirées, a wedding reception for his son, that Louis XV first talked to Jeanne-Antoinette. By 1744 she had become the king's sole mistress and she separated from her husband to live at Versailles as a prominent member of the royal court. She remained the sole mistress of Louis XV for five years, using her influence at court to encourage France's alliance with Austria and, along with her brother, to initiate major public building projects such as the Place de Concord. It was Madame de Pompadour who once said, *"Après nous le deluge!"* prognosticating the terrible bloodshed of the revolution that would follow their golden age.

The courtesan tradition in France survived the revolution and the Napoleonic years, and by the nineteenth century, courtesans had gained wide social respectability, unlike their sisters in the streets. In a sense courtesans were like the supermodels of their day and it was in this climate of popularity and glamour that Sarah Bernhardt, the daughter of a notorious Dutch courtesan (1844–1923), took center stage. Bernhardt was almost a born courtesan but she became an actress, though both professions were regarded as equally scandalous for women at the time. She gained great acclaim in the burlesque theater and eventually became the most famous actress of the nineteenth century. Not only did she act, she painted, sculpted, and wrote—publishing several books and plays. Her long list of lovers included Henri, Prince de Ligne of Belgium (with whom she had a son), Victor Hugo, and Gustave Doré. She moved her stage career to film at the turn of the century, starring in one of the first documentary films of the twentieth century, called *Sarah Bernhardt à Belle-Isle* in 1912.

Bernhardt's career spanned the final years of the grand ladies of the night. By the close of the nineteenth century the long tradition of

European courtesans had ended. Today the lore of the courtesan has devolved into the elite escort services that provide attractive partners to accompany wealthy executives on business trips or social occasions. Although the best of these escorts are educated and well mannered, they are not the refined intellectual courtesans of the Renaissance, a breed whose informed opinions and headstrong, rebellious dispositions not only tickled the fancy of aristocratic gentlemen, but kickstarted the feminist revolution that finally flowered in the suffragist and feminist movements of the twentieth century. Perhaps it shouldn't be forgotten that while some of these courtesans blurred the boundaries between romance and profession, the vast majority of their less fortunate sisters underwent great hardships. Even today the emacipation that courtesans like Theodora and Veronica Franco intitiated is far from universal. Let's hope for a time when women in all professions can, as the activists proclaim, "take back the night."

6

FIRE IN THE NIGHT: NOCTURNAL FESTIVALS— 10 P.M.

And pomp, and feast, and revelry,
With mask, and antique pageantry,
Such sights as youthful poets dream
On summer eves by haunted stream.

JOHN MILTON

A LIGHT COTTON pullover might be comfortable now; it is 10 P.M. and the air temperature has dropped three degrees since sunset. Even though there is a circadian peak in basal body temperature at this time of the evening, it seems a little chilly beneath the clear night sky. The residential sections of the city are almost completely silent, though downtown the night is young and a continuous parade of cars, their polished contours gleaming under the streetlights, are cruising, sightseeing, or searching for parking spots. Some bars have started to reach peak capacity, yet the hot spots—a handful of discos, after-hours clubs, and late bars—are still empty even though they've been open for an hour. Inside the music is loud and bartenders stand ready like bodyguards anxiously awaiting politicians at an inaugural ball.

Outside the city a car whisks down a deserted country highway, its high beams illuminating telephone poles, road signs, and the occasional rabbit scurrying off the road ahead. Insects streak by like shooting stars. A mother is driving her children, asleep in a tangle of legs and arms on the backseat, to their lakeside cottage. They left after sunset and the turnoff to the cottage is only ten minutes away. She is alone with her thoughts and on her car radio a classical station is starting to fade into intermittent reception—Schubert's nocturnes intermingling with the atmospheric static caused by an aurora far to the north. Outside the car window she can see that the stars are bright through the clear country air. The night is even cooler here and mists lie over creeks and ravines. In the small towns she passes, there are people sitting out on their porches.

High above her a jet is gaining altitude. A passenger pulls up the shade to look at the bright stars of the upper atmosphere, the moon reflected from a large lake somewhere far below. Earlier, as the night flight banked over the city after taking off, the passengers craned their necks to look at the immense, glowing network of streets and buildings. There was a fireworks display in a downtown park marking the holiday evening, the intense greens and reds from the tiny explosions looking like jeweled sparks. In the financial district the office towers seemed impossibly small, like an architect's maquette or a miniature train-set landscape before the plane banked once more and the tiny city tilted into the darkness.

Night Festivals

The myriad festivals that take place at night reveal our deep connection to darkness. Many of these are "Eves"—Christmas Eve, All Hallows' Eve, and New Year's Eve—that mark the turning point of the seasons or years. Many, if not most, night celebrations incorporate fire in some way. Perhaps bonfires and fireworks rekindle ancestral memories of our first conquest of darkness deep in prehistory. Water, too, is an element in many night festivals—it seems to take on a special, sacred quality at night. Some nocturnal celebrations are the continuation of daytime festivals, while others, such as the Burning

Mountain Festival in Japan and Great Shiva Night in India, occur only at night and have no associated festival day that precedes or follows them. Of these two festivals perhaps the most purely nocturnal celebration is the Hindu festival of Great Shiva Night (Mahashivaratri) because it is held on the darkest night of the lunar cycle, during the night of the new moon. Mahashivaratri is a sensual festival that takes place in the Hindu month of Phalguna, which corresponds to early March in the Julian calendar. Its theme of earthly pleasure and fertility is evident in the legends about its origin. Hindus believe that this is the night that Shiva first manifested himself in the form of a *linga*, or phallus. But Mahashivaratri is not simply a series of traditional rituals; it is also imbued with a meditative agenda. Many Hindus are convinced that only in the pitch black of a moonless night can the inner light of universal wisdom reveal itself. Out of the darkness, they insist, comes true light.

Other festivals of the night take this notion of light in darkness more literally, and use fire as their central spectacle. It was not all that long ago that fire held magic in its flames, and during a night festival even a torch, a candle, or a lantern can share a portion of the same hypnotic attraction that we continue to have for fire in the night.

Burning Away the Ghosts: Purifying Flames

Although no evil spirits are repulsed by its flames, the Burning Mountain Festival, or Wakakusa Yamayaki, is perhaps the most spectacular of all the nocturnal fire ceremonies in the world. Wakakusa Yamayaki is held on the outskirts of Nara, Japan, the evening of January 15. It is a spring festival, even though it takes place in the heart of the Northern Hemisphere's winter, and the ceremony begins with a Shinto ritual just after sundown. A fireworks display begins the incendiary portion of the evening, and as the pyrotechnics wind down, eighty-one acres of dry grass are set aflame by carefully coordinated fire-starters. In less than ten minutes the whole of Wakakusa mountain is a raging brushfire. This is not a small conflagration. The flames leap more than forty feet high, illuminating the tall plumes of smoke a deep pink and maroon color. Sometimes fiery convection twisters appear inside the fire, their weird, undulating shapes weaving through the flames. The inferno is so large that the winter sky is illuminated a deep red for miles around, particularly on cloudy nights, and the burning mountain can be seen clearly from any part of Nara. Witnesses say that it appears as if a volcano had erupted on the outskirts of the city.

The Burning Mountain Festival may be the most extreme example of a large, ritual fire, though many large fires, called bonfires, are associated with other nighttime celebrations, from the ancient Persian festival of No Ruz, to May Day in modern Sweden. Celebrants sometimes jump through the flames of these special fires to burn away evil spirits and ghosts. Certainly the New Year festival of No Ruz exemplifies this tradition.

The Iranian New Year, or Tahvil, is celebrated around the nineteenth to twenty-first of March, at the time of the spring equinox. It is completely unlike other Islamic festivals that follow the lunar calendar because Tahvil is much older than Islam. In ancient times the Iranian Persians observed the last ten days of their year, which fell on the spring equinox, as a feast of all souls—a time when departed spirits returned to their old homes. In some remote villages special fires are still lit and allowed to burn all night in braziers on housetops just as they once did in pre-Islamic times. These rooftop fires served the opposite function of the ensuing No Ruz fires, as they were meant to welcome the spirits of dead relatives. The Zoroastrians (followers of the original Persian pre-Islamic religion in the region) called this the Suri festival. (The word *suri* means red and fiery.) In some respects the

Suri festival is the seasonal mirror image of the Nordic festival of Winternights that we will encounter in just a few pages.

Today, modern Iranians celebrate Tahvil for thirteen days, but the last Tuesday night of the festival, called No Ruz, is still the nocturnal peak of the observance. Since the advent of Islam, this old Zoroastrian festival has become a secular celebration, though some elements of Islamic traditions have crept into it over the years. For example, celebrants still light large bonfires as the Zoroastrians did, but they do so on the last Tuesday of the year because Wednesdays are considered bad-omen days in Islamic tradition. Otherwise the observance of No Ruz remains much as it was in the past. Many modern Iranians continue to practice the ancient tradition of leaping over the flames of the No Ruz bonfires in order to purify themselves. Even today celebrants shout, "*Sorkhie tu az man, zardieh man az tu,*" as they leap through the flames, which means, "Your fiery red comes to me and my sickly yellow goes to you."

> *The wind is hushed, the starlight pales,*
> *The dismal moon her features veils;*
> *As magic-mad the hosts whiz by,*
> *A myriad sparks spurt forth and fly.*
>
> GOETHE'S *FAUST,*
> "WALPURGIS-NIGHT"

The old Nordic festival of Walpurgisnacht was another bonfire festival. It was held a month after the Persian festival on the night of April 30, the eve of May Day. Nordic legend held that Walpurgisnacht was a "witching night," when witches flew unseen on their diabolical errands. It was also believed that on Walpurgisnacht the dead returned to roam the earth because this was the last night of the Wild Hunt, a retinue of terrible and fierce hunters that rode through the skies led by Odin, who were themselves followed by leagues of the dead. When the hunt ended on Walpurgis, these legions of dead souls flooded the earth for one evening. In that sense Walpurgis had more than a passing resemblance to Halloween or All Souls' Eve, even though it was celebrated in the spring.

At midnight on Walpurgisnacht great bonfires were lit throughout the Nordic countryside. These fires had to be kindled from flint, and not from the embers of hearth fires that had burned continuously through the winter. As with the Persians during No Ruz, it was customary for children to leap over glowing embers or even through the flames of these bonfires in order to "burn the witches" that might be clinging to them invisibly, like hungry ghosts. Fires burned the following night, May Day, as well. Beltane, as May Day was known to the Celts, commemorated the beginning of the time of light, and in the far north, a prelude to the time of the midnight sun. The Druids also celebrated Beltane with large fires, and this tradition continues up to the present in the many variations of May Day fertility celebrations throughout Europe. But even as celebrants danced around the May Day bonfires, dry wood was still being stockpiled for the even greater fires of the midsummer solstice.

The celebration of the summer solstice, when the power of the sun is at its height, was an ancient European tradition that predated even the Druids. Fires burned, and in some cases continue to burn even today, on midsummer nights from northern Europe to the Middle East and from Normandy to Russia. In some countries, such as Sweden, Midsummer's Night remains one of the most important festivals of the whole year, outdoing even Christmas and New Year's Eve. In Sweden huge bonfires (once called Balder's balefires) are lit as soon as darkness descends, and celebrants still dance around the fires and leap through the flames, just as Vikings did during Walpurgis and the Persians did on No Ruz. Another Swedish custom on Midsummer's Night is to throw toadstools into the flames to repel the trolls and the evil spirits that are said to emerge from mountain caves that night. The purifying power of the balefires is believed to be particularly auspicious for lovers who join hands and leap through the flames, for the balefire is said to burn away any curses or evil spirits that might be plaguing the couple.

In Germany, up until the nineteenth century, the white-wine-producing regions of the Mosel River celebrated Midsummer's Night not only with bonfires, but also with large burning wheels. Households in each village would contribute straw to incorporate into a huge wheel that was mounted at the top of the highest local hill. When it was completely dark on Midsummer's Night, the wheel was ignited and two young men would guide the flaming disc as it careened down the hill through the vineyards, showering sparks everywhere. If the burning wheel made it to the river, it was considered a good omen for an abundant wine crop, though I would imagine that during a spell of dry weather the occasional vineyard must have been set alight by the burning disc, completely negating any beneficial effects of the ritual.

There was, in fact, a midsummer fascination with incendiary discs throughout Germany. In Würzburg, during the late Renaissance, it was customary on Midsummer's Night to launch large burning discs of wood from the side of the mountain that overlooked the town. In the night sky the spinning discs left paths of flame that were said to resemble fiery dragons.

'Tis now the very witching time of night.
WILLIAM SHAKESPEARE

Bonfires also burned in northern Europe in the late autumn. The Nordic festival of Winternights, which was held over a five-day period from October 29 to November 2, marked the beginning of winter in the northern lands and was celebrated by wild dancing, drinking, and licentiousness. Winternights was assimilated by the

Celts, though their version of it was a feast called Samhain, celebrated on November 1, which later became Christianized to "All Souls' Day," or "All Hallows' Day," as it was formerly known. The day before All Hallows' Day was called Hallows' Eve, now known as Halloween.

Like the Norse, the Celts regarded the first of November as New Year's Day, and consequently, Samhain was also New Year's Eve, a traditional time of divination and purification. In every home a new fire, not kindled from the old, would be set in the family hearth. The Celts believed that witches flew on brooms through the air on this night and that tabby cats transformed into black horses that witches also rode. Fairies, hobgoblins, and ghosts were thought to be about as well. In Scotland young boys would go from house to house to beg for peat for great fires, called *samhnagan*, usually with the greeting "Ge's a peat t' burn the witches." You can hear today's refrain of "Trick or treat?" rhymed in the old phrase.

After Samhain the winter nights grew longer and colder. The next fire festival in the northern European calendar was Jul, or Yule, as it later became known. In fact, our current celebration of Christmas and Christmas Eve as the preeminent family holiday evolved naturally out of one fiery aspect of this Norse festival—the Yule log. The Yule log was a smaller, domestic version of a festive bonfire. It was necessary because the winter solstice was too cold for the great communal gatherings around the large outdoor bonfires that characterized other Nordic festivals. In the far-flung Norse settlements there were few large buildings for public gatherings, so the Yule fire of renewal and hope was lit individually, in the hearth of every home, and thus the festival was marked at the outset by a familial privacy we can still recognize in today's Christmas. (In Westphalia, Germany, Yule logs were extinguished before they burned completely and were kept near the fireplace in order to be relit in case of a thunderstorm. It was believed that a house in which a Yule log smoldered could never be struck by lightning. This custom was also practiced in some parts of France and England.)

The significance of Yule for the Norse, a culture fascinated by fate and absolutes, was that it heralded both an end and a beginning. Even though it was after their celebration of the new year on November 1, the longest night in December marked the end of the journey into darkness and the beginning of longer days, and with them, the dawn of hope. According to Norse legends Odin rode his magic, eight-legged horse, Sleipnir, on this night and it was traditional for Norse children to leave offerings of hay and sugar for Sleipnir in their boots and shoes, placing them by the hearth on solstice eve. They hoped that Odin would leave them a gift in return for feeding his horse and thus the reciprocal tradition of Christmas gifts began—a cookie and milk for Santa (hay and sugar for Sleipnir) and gifts from Santa (read Odin)—as well as the practice of hanging stockings by the fireplace in which burned the Yule log. Deep in the night of Christmas Eve, in the lights that adorn today's Christmas trees and homes, is the Norse fire of midwinter's hope. Halfway around the world, in Japan, the fire that signals the birth of spring occurs later in the season, though it is equally sacred.

Fire and Water

The Burning Mountain Festival is perhaps the most extreme of the Japanese fire celebrations, but the Omizutori Torch Festival is almost as spectacular, though in a subtler way. The Torch Festival is held in March, in a Shinto temple called Nigatsudo Hall, just outside the town of Nara. For two weeks prior to the Omizutori ceremonial procession which crowns the Torch Festival, ten huge burning torches, each of which measures over eighteen feet tall, are kept lit all day and night in front of Nigatsudo Hall. On the night of

March 12, the eve of the ceremony, eleven more torches, each of which stretches a colossal twenty-four feet high, are lit. These torches burn for some time until the the fire ceremony itself, called Dattan, takes place.

Dattan is a deep-night ritual that starts very late: 2 A.M. on March 13, and around 1 A.M. on March 14. Walking in single file, a line of priests from Nigatsudo Hall emerges through the temple doors carrying large pine torches. They circle the veranda of the temple in a procession before they file to the temple's sacred well to draw water. Here the Omizutori ceremony becomes a fire-and-water ritual, where sacred spring water, doubly charged with the primal spirit of earth as well as being infused with the vital energy of vernal darkness, is raised into the fiery light of the torches.

The theme of fire and water is perhaps most exemplified by the Thai festival of Loi Krathong (also called Yi Peng in northern Thailand), which takes place on three nights during the full moon of the twelfth lunar month (which usually falls in November). Here fire, in the form of candles set adrift on the water, literally meets water. Loi Krathong is essentially a nocturnal river-worship festival, in this case honoring the spirit of the Ping River, which flows through central Thailand.

When dusk falls on each night of the festival, celebrants light lanterns in front of their homes and then join a parade down to the river, where they launch floating lanterns and small rafts adorned with candles and offerings out into the river. The cumulative effect of many hundreds of these lights flowing into the current is one of an incandescent, glowing band of multicolored light that follows every meander of the great river downstream. At the same time as the floating lanterns are launched, people release small hot-air balloons, also carrying little lanterns, into the sky above the river. These lights drift through the night with the wind, and after rising above the treetops, usually set off in formation, shepherded by the prevailing wind.

On the final night of the festival the winner of a local beauty contest is carried to the river at the center of a parade where she, too, is placed on a raft adorned with lanterns that is launched onto the river. The notion of sacrifice underlays this aspect of the festival, though today the offering of a beautiful young woman has become a mere reflection of the ancient tradition.

The Stroke of Midnight

O nights and feasts of the gods!
HORACE

Many of the previously described nighttime festivals are New Year's celebrations of sorts, though they were not as dependent on the exact time of midnight as New Year's celebrations have become since the advent of the Julian calendar and accurate clocks. In today's world the two dominant New Year's celebrations are timed to the second, though for the Chinese it is the most riotous, the most festive, and the most important of all their festivals. It is also the most familial. As in other New Year's festivals, the celebrants stay up drinking and watching special shows on television. But the Chinese observance of the new year has one very poetic ritual that is unique. At the exact stroke of midnight the Chinese fling open all the windows and doors of their houses in order to release the old year. It is this moment of opening the home to the night air that is perhaps the most magical moment of the entire evening, for with the fresh midnight air comes hope for the new year and, perhaps, a first, virginal whiff of spring.

By far the majority of the rest of the world's inhabitants celebrate New Year's Eve on December 31, a date that originated soon after Julius Caesar rearranged the Roman calendar to make it more accurate. It's hard to believe the accomplishments that Julius Caesar stuffed into his fifty-six years on earth—besides conquering Britain, Germany, and Gaul in a series of brilliant military campaigns, he became the first absolute emperor of Rome, not to mention the lover of Cleopatra. In his capacity as supreme ruler, no detail was too insignificant for his attention, and as soon as certain calendrical discrepancies were pointed out to him, he dispatched them in his usual compulsive manner. He took time into his own hands, so to speak, lengthening the year to three hundred and sixty-five days from three hundred and sixty, and reestablishing January as the first month of the year instead of March.

Although January 1 had been designated as the start of the new year a century before Julius Caesar remade the Roman calender, it wasn't until his calendrical reforms that the date became firmly established. According to the new Julian calendar, the very first New Year's Eve was on December 31, 44 B.C. The name *January* was chosen because it referred to the Roman god Janus, who had two faces—like *Janus* the first month of the year had two faces, facing back at the old year and forward to the new one. In a deeper sense this began the dual tradition of nostalgia and hope that has marked the Western celebration of New Year's Eve ever since.

The Romans celebrated the transition from one year to the next at the precise moment it occurred, 12 A.M. New Year's Day, as does much of the world now, due to the global adoption of a slightly modified version of the Julian calendar called the Gregorian calendar. But this wasn't always so. The Christian church refused to recognize January 1 as New Year's Day (being a pagan Roman festival) until the seventeenth century; up to that time New Year's Eve was officially observed on March 24. (Jews still celebrate their new year, Rosh Hashanah, during September or early October, and the Muslim new year follows the Muslim lunar calendar. The Chinese new year also falls outside the Julian norm, being celebrated in late January or early February.) For most of the Western world, New Year's Eve is the only

completely secular night festival besides Halloween, and it is celebrated with the same excess and revelry as was the Roman Saturnalia.

New Yorkers crowd Times Square at midnight while Neapoli-
tans throw dishes and pots out their window to bring good luck.
Similarly Muscovites toss empty vodka bottles above their heads at
the stroke of midnight, draining the old year and symbolically
smashing it on the cobbles of Red Square—though some stray
bottles must occasionally compound the inevitable hangovers of
New Year's morning by striking the heads of other drunken
revelers.

Symbolic colors are often worn on New Year's Eve: Peruvians wear
yellow underwear to usher in the new year, while Italians wear
something red for good luck. In Rio de Janeiro, thousands of
Brazilians wearing white clothes run noisily into the ocean at midnight, carrying offerings of candles, sugarcane, tobacco, alcohol, and
flowers for the goddess of the ocean, Iemanja. For Danes it seems that
every year is a leap year: they hop off chairs at midnight in order to
jump into the new year.

When I was growing up we used to celebrate New Year's Eve with
clam chowder and new year predictions. My mother would melt lead
in an old saucepan on the kitchen stove and then pour the molten
metal into water, creating surrealistic shapes of congealed lead. The
rest of us would interpret these shapes as prognostications for the year
to come, delivering grandiose prophecies as well as more considered
predictions that would form the basis of wagers. TV served as the
countdown clock, and as Guy Lombardo began "Auld Lang Syne," I
would go out onto the back porch and listen to the new year arrive in
the relative silence of the outdoors—the distant shouts and singing of
revelers interspersed with the explosions of distant fireworks in the
cold stillness of deep, winter night that, itself, seemed unperturbed in
any way by our earthly commotions.

Fire in the Sky

The visual extravagance of fireworks is the perfect nocturnal complement to national holidays, be they anniversaries of independence or the birthdays of great monarchs and politicians, and a dark night sky is an ideal backdrop for their brilliant, pyrotechnical colors. There is something of the grandiose tribute in fireworks, like victory bouquets of fire, though some ethnologists have argued that the spring firework ceremonies of European nations are remnants of the spring and solstice bonfires that once burned from Norway to Iran. In Canada, Victoria Day (Queen Victoria's birthday) is celebrated on May 21, and on that night families from my neighborhood meet in our local park to light fireworks. It's an ad hoc affair of parents and children, with each family contributing a few Roman candles, some "sonic boomers," and a burning schoolhouse or a Catherine wheel. At twilight the parents arrive and begin setting up sand buckets to hold fireworks while children scream and chase each other with sparklers and grandparents unfold lawn chairs. (There is something about the preparations of the fireworks themselves, a task that is almost always allocated to the fathers, that has somewhat of a military feel. After all, the borderline between exploding artillery bombs and a large fireworks display is somewhat blurry.) The youngest children are most susceptible to the holiday excitement—just being out after dark on a warm spring night with neighborhood friends is thrilling in and of itself. Then, without fanfare, one family lights their first Roman candle and the evening begins. Soon several displays are going, and everyone stands mesmerized by the colors of the chemical fires, detonating like miracles against the dark sky. But local fireworks pale in comparison to the huge public displays mounted by licensed pyrotechnicians.

For several years Toronto has hosted an international fireworks competition that takes place in early July. These displays are always state-of-the-art and always monumental. They are set off from barges that float in the harbor in front of massive crowds watching on the shore. During the competition I contacted the festival publicity department and told them that I'd like to interview one of the pyrotechnicians. The next morning I got a call from Enzo Bertolini, the director of a pyrotechnical team. He told me to come down to his hotel near the waterfront and that he would escort me on a tour of the fireworks platform that afternoon.

Enzo was waiting for me in the lobby. He was tall, with more than a passing resemblance to film actor Jeff Goldblum, though he had shoulder-length blond hair. Enzo smiled at me and shrugged as if to excuse the tiny, iridescent-blue cell phone clamped to his right ear. Continuing to talk, he reached out and shook my hand and then gestured at me to accompany him outside. (Later, I reflected, it seemed that most of the time we were together he was involved in three-way communication. It seemed quite natural for him.) He was still talking when we got into his car, though he ended his call when he turned the key in the ignition. In perfect English he asked about my research and what I already knew about fireworks. I told him what I'd discovered through my rudimentary investigations, that fireworks were invented in China as a consequence of the invention of gunpowder, and that gunpowder was a mixture of potassium, nitrate, sulfur, and charcoal. The first Chinese fireworks displays, I went on, probably occurred around A.D. 900 (during the Tang Dynasty) and then spread along the trade routes westward. "So far, so good," he said, "but do you know who the first Europeans were to use fireworks?" "No," I replied. "The Italians!" he said. "The first Italian display was in Florence, in the fourteenth century." Enzo's phone rang during our historical synopsis and he was talking as we reached the waterfront and parked beside a quay with a couple of big boats moored alongside.

Still talking into his cell phone, Enzo walked to the side of the quay and motioned toward a handrail and ladder leading over the side.

Moored at the bottom was a Zodiac, the same type of inflatable boat that Greenpeace activists used to harass whaling ships. Somehow, with the phone at his ear, Enzo managed to climb down the ladder, unlatch the mooring lines, and start the engine. I practically fell off my seat when he accelerated out of the harbor and across the bay. Soon we were at the fireworks barge, which, at first glance, looked like a small industrial island. Enzo ended his second call and docked the Zodiac. Then he took me on a tour of the barge, a floating islet of pyrotechnic munitions moored in the middle of the harbor.

He told me that he and two assistants had spent the best part of two days installing the canisters, fuses, and wiring for tonight's display. The complex network of wires, junction boxes, and fuses ran like spaghetti all over the surface of the floating launch platform, snaking between dozens of cylindrical shapes—the launchers themselves—draped in plastic sheeting. It was hard not to step on them, though they were arranged in rows so the technicians had access to them. Enzo said the barge had to be turned according to the prevailing wind on the night of the display because sparks from the launchers could ignite adjacent fireworks. The rows of fireworks had to be lit not only by the evening's firing order, but also with an eye to wind direction and the safety of the pyrotechnicians. Usually one or two assistants man the barge in case anything goes wrong. This is a hazardous job, and pyrotechnicians can lose their lives if a faulty charge ignites the rest of the fireworks.

I didn't see any of the tall, long shapes I expected. Everything seemed short and squat. Where are the rockets? I asked him. He smiled and replied that fireworks displays no longer use rockets; they use mortars that fire ballistic packages of timed explosives called "shots" two hundred to three hundred feet high. He went on to explain that the two other factors in fireworks are the height at which they explode, which is determined by the length of the fuse and the amount of gunpowder used to propel the shot upward, and the spread of the firework, or how wide it is at its greatest extent in the sky. In classic fireworks there has to be a balance between the height and the spread, but the most recent techniques apparently use wide spreads at lower heights.

Enzo told me about recent developments in the art of pyrotechnics. Contemporary displays are the products of hundreds of years of fine-tuning. Tonight's display would be coordinated with music, in this case, Handel's *Royal Fireworks Music.* The timing of a musically synchronized display couldn't just be approximate, he insisted, it had to be perfect, especially during an international competition like this. No matter that he was dealing with enough explosive force to blow the barge right out of the harbor, he also had to ensure that every millimeter of fuse, every ignition circuit worked flawlessly. The art of fireworks, Enzo pontificated, is complex; it combines ballistics, trajectories, timing, and explosive yield. Enzo uses a laptop computer to program and coordinate the electronic ignition sequences of the firework fuses.

After our tour Enzo took me back to the shore and dropped me off. He had to return to the barge to prepare for the evening. That night I came to watch his display from the harborfront. I was a little late and unprepared for the crowds; thousands of people had assembled. I had a special pass for the main viewing stands but I wasn't able to squeeze through the crowd before the fireworks started with a series of tremendous, ear-rattling percussions. Then the music began, from big speakers near the main viewing stands—Handel's fireworks symphony played as it was meant to be played, in the summer night, as the master-pyrotechnicians wove their magic tapestry of fire in the blackness.

The fireworks began with a colonnade of silver fountains of sparks that geysered up one hundred feet into the sky, mounting higher and higher until, suddenly, with a series of big bangs, cherry-red star-bursts of color capped each of the fountains. The red stars turned lime green and suddenly, for a moment, we were looking at a row of brilliant, iridescent palm trees, a vision of tropical paradise. The fireworks were like a fantastic music video, and Enzo's coordination of Handel's score to visual shapes and astonishing colors was extraordinary. The show kept mounting, impossibly, with each explosion and effect a little bigger and brighter than the last. It was hypnotic and exhilarating at once, as the fireworks blossomed like hallucinations in the night sky. Each star-burst would expand out of nothingness like a small universe of color, then droop and fade, its sparkling jets spreading to fill the night with neon-bright streamers. The entire display was punctuated by percussions and bangs that sometimes completely drowned out the music. Then, as the display built to a climax, a series of intense purple clusters were detonated, followed by small sparkling green twinkling explosions that sounded like ballistic popcorn. After that, four tremendous bangs, a giant star-burst of unearthly deep blue that seemed to fill the whole sky, and it was suddenly all over. Only the smoke drifting above the slowly dissipating crowd and the memory of those otherworldly colors against the night sky remained.

7

THE NIGHT WITHIN: THE BODY AT NIGHT—11 P.M.

I T IS 11 P.M. AND a great many people are asleep. The air is beginning to clear as the day's pollution starts to settle. Throughout the city a long silence is rising like a quiet tide, leaving only a few promontories of noise—the highway maintenance workers laying hot asphalt on the expressway under arc lights, the noisy restaurants and bars in the entertainment district, a siren wending its way toward an emergency ward. Among those who are awake, a few will be sneezing: our sensitivity to dust peaks around 11 P.M.

In the country the songs of field crickets are slowing down as the night air cools. If you don't have a thermometer, you can estimate the evening's temperature by listening to the chirp rate of the snowy tree cricket: adding forty to the number of chirps in thirteen seconds gives a fairly accurate reading of the temperature in degrees Fahrenheit. The whippoorwill's insistent, hollow refrain sounds across the crickety fields. There is a nervous boldness in its song, as if it could stop time somehow by repetition of the same phrase.

Night and the Senses

How silver-sweet sound lovers' tongues by night,
Like softest music to attending ears!
WILLIAM SHAKESPEARE

At night, as our visual sense is diminished our other senses become more acute. The senses of touch, smell, and particularly of hearing are all increased in a temporary amplification that is similar to the way these senses are permanently heightened in people who are blind. There is a telling description of this sensory augmentation effect in Tim O'Brien's book on Vietnam, *The Things They Carried*. At one point he recounts how his sensitized hearing wreaked havoc with his war nerves during a night watch in the Vietnamese jungle. Sitting, waiting for a possible night attack, he describes how, "as the night deepens, you feel a funny buzzing in your ears. Tiny sounds get heightened and distorted. The crickets talk in code: the night takes on a weird electronic tingle." Other, earlier writers have noticed this same effect. Edgar Allan Poe wrote, "Sound loves to revel in a summer night," and Victorian poet Edmund Hamilton Sears declared that our nocturnal hearing becomes so acute that, if we listen carefully enough, we can hear divine music: "Calm on the listening ear of night / Come Heaven's melodious strains . . ."

Our senses of smell and taste are also heightened in the dark. In fact, a new trend in restaurants in Europe enlists darkness to supposedly enhance the gastronomic pleasure of their diners. In Zurich, a restaurant called Blinde Kuh (blind cow) has a long waiting list for customers who wish to eat their gourmet fare in total darkness. Blind or visually impaired waiters guide the customers to their tables. The trend is spreading and some restauranteurs in the Unites States are looking into transplanting the idea across the Atlantic. Chic restaurants have always employed dim lighting, though the novelty of eating in darkness will likely never supplant our need to see what we are eating.

Night magnifies our sense of touch as well. We've all felt for a light switch in a dark room, and the thrill of running our hands along the smooth contours of another's warm skin seems marvelously intensified by darkness. Night is a tactile realm, a kingdom of touch. It seems almost as if night itself were graspable, as Walt Whitman declared in *Leaves of Grass*, "Press close bare-bosom'd night! Press close magnetic nourishing night! / Night of south winds! night of the large few stars! / Still nodding night! Mad naked summer night!"

The Circadian Rhythm

It is always night inside the body. Bright sunlight passes through the skin and on into deeper tissues to a depth of approximately three centimeters, but past that it is pitch dark. The perpetual night of the body is where each of us began, and so we are truly children of the physical night. We quickened in the warm, aquatic twilight of our mother's womb. Even if none of us seems to remember it, our first nine months were spent in darkness, with only occasional, faint glimpses of light to tantalize us. (Pediatricians report that in the last few months of pregnancy the eyes of fetuses are developed enough that they can see a dim red glow through their mother's belly if she is lying in the sun.)

In our heads it is even darker; a little more than two centimeters inside the cave of the skull it is always black. Raw, unprocessed light never touches the brain. The brain prefers to receive its light digitally, in the form of discrete impulses carried by nerves from the eyes into the cryptic darkness of the brain. In our minds, daylight is only represented; it never actually shines there. The information that our eyes carry to our brains from the outside world is also crucial for maintaining the daily rhythms of our bodies. It is through our eyes that our biological clock, also known as the circadian clock, is set —the word *circadian* comes from the Latin *circa* ("about") and *diem* ("a day"). Light, or the lack of light, shining on special ganglion cells in the retina that are tuned to detect brightness and darkness causes them to transmit their impulses to a part of the brain called the suprachiasmatic nucleus (SCN). This small group of neurons deep in the brain is the mainspring of our circadian clock, as they control the release of various hormones and secretions that regulate our daily cycle of sleeping and waking.

The circadian rhythm of alternating darkness and light was present at the very origins of life on our planet and it is the indivisible binary pulse at the center of our existence. In fact, every living cell of our bodies is affected by the interplay of day and night. There are many other circadian cycles that operate outside the influence of our main biological clock—circadian rhythms are programmed into more than thirty of our genes, and some of our organs, like the liver, have their own twenty-four-hour cycles. Not all these rhythms are synchro-nized—the liver may be awake while the body sleeps, and vice versa. We rarely experience the effects of any of these subcycles, but we are very much aware of the dominant circadian rhythm that is controlled by the SCN.

When night falls, a cascade of hormones is triggered by the SCN. These secretions flood the body and initiate a complex series of transformations. Around 9 P.M., a few hours after dark, the SCN signals the pineal gland (a cone-shaped organ in the center of the brain once thought to be the seat of consciousness) to begin melatonin secretion. During the day, melatonin levels are so low they are undetectable, but a couple of hours after dark, triggered by impulses from our circadian clock, melatonin floods the body like a velvet tide. Melatonin has a powerfully sedative effect, and doses measuring a fraction of a milligram can induce sleep. It can also shift the sleep cycle for those suffering from jet lag and has recently been claimed as an anti-aging substance, though there is no real evidence to support that claim. Melatonin levels peak between 1 and 2 A.M.

At the same time as melatonin is preparing our body for sleep, aldosterone levels are also climbing. Aldosterone is a nocturnal hormone secreted by the adrenal gland that regulates the levels of potassium and sodium in the bloodstream at night. Regulation is necessary because we do not urinate as frequently at night, so our blood serum levels have to be maintained by aldosterone. Although, as we will find out later, it is not as powerful a suppressant as the hormone cortisol, aldosterone does suppress the immunosystem. At around 10:30 P.M., the rhythmic contractions of the intestines and colon (peristalsis) are slowed in preparation for sleep.

Melatonin causes us to fall asleep between the statistical norm of 11 to 12 P.M. on a normal, workday night. At the same time, our core body temperature begins to drop, particularly rapidly after we fall asleep, reaching its lowest point, 96 degrees Fahrenheit (a full two and a half degrees lower than our normal 98.6 degrees Fahrenheit), at around 5 A.M.. It is at this same time that cortisol, a hormone that elevates the blood sugar levels, begins to rise. Cortisol is released by the adrenal glands when the pituitary hormone ACTH stimulates them. Cortisol also triggers the release of fatty acids from adipose tissue and amino acids from muscle tissue—the liver then converts these amino acids into glucose for use by the brain while the fatty acids provide direct fuel for the skeletal muscles. These responses are all stress adaptations that allow the body to react to trauma, be it physical or psychological. Cortisol ebbs and flows in accordance with our circadian clock—its levels in the bloodstream reach their lowest point around midnight and peak in the last hours of sleep, between 5 and 7 A.M. Like a self-manufactured caffeine, cortisol appears to prepare the body for waking.

The usefulness, and toxicity, of cortisol (not to be confused with the medication cortisone) is a topic hotly debated by endocrinologists. Because it is secreted naturally by the body, it has been presumed to be a benign hormone, but recent evidence seems to be pointing in the opposite direction. In the September 1997 issue of *Psychoneuroendo-crinology*, Alfred T. Sapse, a renowned immunologist, issued a cortisol alert. In his article he explained how he had come to believe that cortisol played an insidious role in the development of AIDS, multiple sclerosis, and many other serious diseases. He wrote, "Cortisol is probably one of the most violent immunodepressants there is." The tide of cortisol that floods our bodies in the last few hours of night depresses our immune systems, leaving us unguarded against viruses and disease. If you think your cold seems worse in the morning, you're right: it is.

Poppies of Somnus:
The Physiology of Sleep and Dreaming

> *Sleep falls on forest and field.*
> *See! sleep hath fallen: the trees are asleep:*
> *The night is come. The land is wrapt in sleep.*
>
> ROBERT BRIDGES

Almost all animals sleep. Dogs sleep soundly, even if sometimes their paws and whiskers twitch as they chase dream squirrels. Cats are famous for their eponymous naps, though their short snoozes add up to more than twelve hours a day, almost twice the average sleeping time for a human. Until recently it was believed that sharks couldn't sleep because they rely on forward motion to oxygenate their gills—if they stop swimming they die—but many species of sharks have now been documented sleeping on the sandy floors of underwater caves where there are enough ocean currents to aerate their gills while they rest. I have seen films of scuba divers stroking nurse sharks when their great unblinking, lidless eyes were faintly opaque with sleep.

Ants sleep, or at least go into stasis. My son had an ant farm when he was young and he noticed that his ants would sometimes find a quiet spot, often a small unused side tunnel within the nest, and remain motionless for several hours. A form of sleep perhaps.

Porpoises sleep submerged, though they wake up every few minutes to surface and breathe. Sheep and cattle slumber very little, although horses need about seven hours of sleep a day—as well, they are one of the few mammals that can doze while standing. Otherwise most warm-blooded animals need to lie down or curl up while sleeping, except for birds, most of which sleep while perched. The European swift outdoes all other birds, however—it can sleep while flying; how zoologists established this fact, short of an interview, is beyond me. In the same vein, interestingly enough, there *are* many verified accounts of soldiers sleeping while they march. My cousins claimed that their father, my uncle, used to read the newspaper and doze while he drove to his cottage!

How often have we heard the phrase "We sleep away a third of our lives," which reveals a common notion that sleep is a null void of downtime. Worse than that, sleep has been historically regarded as a kind of death, a fact recognized by Hesiod in the eighth century B.C. when he wrote, "Night, having Sleep, the brother of Death," and by Samuel Daniel, who commented on our ambivalence toward sleep in his sixteenth-century verse, "Care-charmer Sleep, son of the sable Night, / Brother to Death, in silent darkness born." Perhaps our fear of night is grounded in this association, particularly today, as the technologically driven pace of our present lives completely reinforces the notion that being unconscious is like being dead. It's no coincidence that it was the inventor of the electric lightbulb, Thomas Edison, who declared that sleep was "a waste of time." He was a hyperactive insomniac, however, so his prescription for alertness was inappropriate for the vast majority of the population. In some respects his legacy has been a curse, because as a result of his invention, the twentieth century witnessed a burgeoning epidemic of sleep disorders related to shift work and sleep deprivation. Sleep may have become a "waste of time" for our adrenalized civilization but, biologically speaking, it is still time well wasted.

Despite decades of research, sleep remains a mystery. Dr. William Dement, director of the Stanford University Sleep Disorders Research Center, writes: "There are theories, guesses, some very suggestive experiments, but nothing has yet pinned down the precise reason the body must lie still and the mind draw inward for a third of every day." No one would contest the claim that sleep is restorative in some way. Shakespeare wrote that it was the "balm of hurt minds" as well as "the chief nourisher in life's feast," and it is known that the human growth hormone GH is released by the pituitary gland during sleep, repairing tissues and strengthening the immune system despite the presence of cortisol. Perhaps "Sleep is the sovereign remedy," as my father used to say, and indeed sleep is an essential period of regeneration for our bodies and minds. Yet it is not as therapeutic as many believe. Death also stalks the early-morning hours. Many sudden heart-attack deaths occur between 3 and 4 A.M. because that is when the irregular heart rhythms that accompany deep sleep are most frequent. (A friend of mine suggested that anyone worried about dying in their sleep at this time should set their alarm for 3 A.M. and stay up for an hour.) As well, heart attacks caused by sleep apnea (a restriction of the airway that causes snoring sleepers to stop breathing for short periods of time) also strike in the early-morning hours. Between 5 and 7 A.M., when cortisol levels are at their highest, there is yet another peak in nocturnal deaths. Night nurses report that the fevers of sick patients often peak at night, though, on the positive side of the ledger, they also report that the most frequent time for the onset of labor is between 1 A.M. and 7 A.M., though why this is so remains a mystery. At night it seems the stork rides shotgun with the grim reaper.

Sleep researchers have also discovered that how long we sleep every night is definitely related to how long we will live. By all means get your beauty sleep, but not too much, as lingering in bed could be lethal. In the February 15, 2002, issue of *Archives of General Psychiatry*, the published results of a study of more than a million people by

Daniel F. Kripke of the University of California found that those who slept more than eight hours a night had a higher mortality rate than those who slept less. The highest longevity was found among those who slept between 6.3 and 7.4 hours. This is good news for those who think they're not getting enough sleep. Statistics show that the average American sleeps 6.5 hours a night, which is close to the center of Kripke's designated sweet spot of healthy slumber. Sleep researchers have also discovered that there is a definite relationship between sleep and heart ailments. People who sleep less than the optimal amount seem to have a higher statistical incidence of angina pectoris heart pains, though those who sleep longer have more heart attacks.

Modern sleep research began with Dr. Nathaniel Kleitman's "discovery" of rapid eye movement (REM) sleep in the early 1950s. Kleitman wasn't working in a vacuum though, as the first recorded observation of rapid eye movements was made more than two thousand years ago by Aristotle, who believed dreams were "the life of the soul." Aristotle noticed that the eyelids of sleepers twitched and pulsed while they slept and correctly deduced that these movements were caused by the eyes moving beneath the eyelids as the sleepers dreamed. But it was left to Dr. Kleitman, a professor of physiology at the University of Chicago, to prove the link between rapid eye movements and dreams by designing the first experiments that began to unravel the mysteries of slumber. He had already observed that the eyes of sleeping test subjects were sometimes active behind their closed lids but he hadn't framed his observations within the context of a scientific study. He delegated this task to one of his graduate students at the time, Eugene Aserinsky.

Aserinsky used his eight-year-old son as his subject and an unused room in the physiology department as his laboratory. He connected his son to an electroencephalograph (or EEG, a brain-wave recorder) and taped electrodes to his eyelids each night to record eye movements. Within days Aserinsky noticed that his son's eyes would sometimes move around "as if he was awake." He also noted how regular these periods of rapid eye movement were. The movements would "disappear for a time, then reappear," he recalled during a recent interview. This regularity fascinated him. It seemed that sleep involved a series of regular and discrete cycles marked by periods of rapid eye movements. Aserinsky documented his sleeping son's eye movements for two years before confirming what Dr. Kleitman had initially guessed.

Out of this and subsequent research, a clearer picture emerged of what our brains and bodies undergo every night as we doze. The process of initiating sleep involves a complex choreography between several regions of the brain. Although, as the old saying has it, sleeping is as easy as falling off a log, falling asleep, at least from our body's point of view, is actually an active process where the brain induces sleep in a well-coordinated series of stages.

When we lie down at night, several areas in the brain have already been activated by the presence of melatonin. Among these areas is a group of cells in the raphe nucleus, located in the upper brain stem, which begin to secrete higher than usual amounts of a neurotransmitter called serotonin. As we begin to fall asleep, serotonin spreads along the neurons of the brain stem to the thalamus, a structure deep in the brain that acts like a switching station, coordinating much of the activity of the highest part of the brain, the cortex, as well as relaying thought impulses to the inner brain and to the rest of the body through the brain stem. The elevated levels of serotonin signal the thalamus to slow down the activity of the cortex, and as a result our thinking becomes less focused and more associative. Researchers can calibrate this deceleration of brain activity by measuring the frequency of brain waves.

Brain waves are the electrical activity produced by neurons in the brain. Because the cortex of the brain is made up of billions of neurons, these neurons have a tendency to harmonize their discharges in electric waves of excitation. These waves (the synchronized pulses of millions of neurons) can be measured by an EEG and have several different modes, or frequencies, depending on if we are working, resting, or sleeping. When we are alert and active, our neurons are firing at their fastest, approximately fifteen to forty cycles a second. This level is referred to as beta. The next lower level of activity, which corresponds to a state of calm reflection, is known as the alpha rhythm, and is in a frequency range of nine to fourteen cycles a second. This is the state we are in when we have tucked ourselves into our bed and have turned off the light. The next lower level, the edge of sleep, or half sleep, is called theta. Here brain waves are slower still, between five and eight cycles a second. The theta rhythm corresponds to the brain wave frequency of the thalamus, which as the keen reader will remember, orchestrates much of the activity of the higher brain. So it is here, on the edge of sleep, that the rhythms of the thalamus begin to flood into the rest of the brain, submerging the cortex in the slow waves of this essential but evolutionarily more ancient part of our minds.

But with the onset of dreams several new things occur in the brain. Serotonin levels in the pons, a structure within the upper brain stem, drop precipitously as levels of another neurohormone, acetycholine, rise. The acetycholine-flooded pons then causes the forebrain, the area of the mind involved with planning and decision making, to shut down almost entirely. It is now that dreams begin their shadowy cinema behind the pulsing eyelids of the sleeper.

Perchance to Dream

The sleepers are very beautiful as they lie unclothed,
They flow hand in hand over the whole earth, from east to west, as they lie
unclothed,

The Asiatic and African are hand in hand—the European and American are
*   hand in hand,*
Learn'd and unlearn'd are hand in hand, and male and female are hand in
*   hand,*
The bare arm of the girl crosses the bare breast of her lover—they press close
*   without lust—his lips press her neck,*
The father holds his grown or ungrown son in his arms with measureless love,
*   and the son holds the father in his arms with measureless love,*
The white hair of the mother shines on the white wrist of the daughter

WALT WHITMAN

Sleep researchers divide sleep into four stages. Stage one corresponds to the lower end of the theta rhythm, stage two corresponds to the upper frequencies of an even slower rhythm than the theta, the delta rhythm. Stages three and four, the deepest levels of sleep, correspond to the slowest frequencies of the delta rhythm. A sleeper's passage through stage one down to four and then back up to stage two is referred to as a single "cycle," which lasts approximately ninety minutes. In the lower, or slowest, delta rhythm we drift into our deepest, dreamless sleep. At this stage our brain wave frequency rarely exceeds three cycles a second. Normally we don't enter this fourth stage of deep sleep until we've been asleep for at least an hour. Then, after passing some time in stage-four deep sleep, our brain waves begin to speed up again until they approach the upper theta rhythm, or wakefulness. It is at this point that we begin to dream.

Dreams, it turns out, come in two modes: phasic and tonic. In the phasic mode the dreamer's eyes show the characteristic REM activity that usually accompanies dreams, but in the tonic mode there are no eye movements, even though there are dreams. Freud conjectured that dreams were the brain's method of keeping us asleep, giving us a decoy reality to engage our attention, and sleep research seems to bear his theory out. Our most intense REM periods, in which we dream most actively, are always in stage-one sleep, just at the edge of waking consciousness before we reach the upper levels of the theta rhythm.

An average night's sleep usually consists of four or five ninety-minute cycles and each successive cycle plunges the sleeper a little deeper into stage-four sleep than its predecessor. Some sleep researchers have conjectured that our vivid dream life might be a recent evolutionary adaptation to civilization. They speculate that our ninety-minute sleep cycles might be the remnants, like behavioral fossils, of the shorter sleeps maintained by our ancestors, who had to wake up several times a night to keep a lookout for predators or invaders.

There is another evolutionary adaptation that has protected sleeping humans over the millennia. Just before we begin to dream, a group of neurons in the brain stem disconnects our muscular control so that we are effectively paralyzed. This paralysis is called atonia and it is a necessary precaution against acting out our dreams. We've all had firsthand experience with atonia—those claustrophobic dreams where we are incapable of moving, where, no matter how hard we try to run, it seems as if we're struggling through molasses, or where we are pinned by some nightmare foe yet cannot yell for help because we cannot speak—all these predicaments are symptomatic of our conscious mind (or at least the remnants of our conscious mind) fighting the insidious paralysis of atonia. The suffocating anxiety of these dreams arises directly from our semiconscious sense that we have become paralyzed as we doze off.

In certain cases, though, the sense of paralysis that accompanies atonia is more than a vague unease that haunts people's dreams: it is a real, waking nightmare. Some people suffer from sleep paralysis, a malfunction of atonia where their body continues to remain paralyzed even after their mind is awake. Sixty percent of us have had at least one episode of sleep paralysis, but for some it is a regular and unwelcome occurrence. For those who are afflicted by this disorder, which statistically makes up more than 2 percent of sleep disorder cases worldwide, it can be a terrifying condition. The American singer Sheryl Crow is perhaps the most famous victim of sleep paralysis. She said in an interview with *Rolling Stone* in 1996, "There would be nights where I would be afraid to go to sleep." She also related how frightening the experience was, saying, "In sleep paralysis, sometimes you get to the point where you are sure you're going to die."

Fortunately, for most of us, atonia shuts off automatically when we awake. But there is another, even more unsettling phenomenon that often occurs just when we're drifting off to sleep, and that is the sudden sense that we are falling or being startled awake. Everyone has probably experienced this plummeting sensation just as they begin to sleep, particularly the automatic jerk that wakes them up. There are many myths surrounding this phenomenon. Folk legend has it that if you don't wake up when you begin to plummet, you will die when you hit the "ground." In reality this sensation of falling and then jerking awake is called a "myoclonic jerk" and consists of the sudden, involuntary, and simultaneous contraction of large body muscles that often occurs just when atonia is cutting in. The cause of myoclonic jerks is not known, though some physiologists conjecture that they may be an involuntary survival mechanism where an electrical impulse from the brain is sent in response to the quick reduction in heart and breathing rates that comes with the onset of sleep, which the brain misinterprets as dying. So there is some truth to the folk belief after all.

Carl Sagan, in his book *Dragons of Eden*, had a much more interesting explanation for the myoclonic jerk, and one that has a certain compelling logic. He speculated that when our prehuman ancestors slept in trees at night, they evolved a safety mechanism that made sure they didn't fall when they went to sleep. The safety mechanism was the myoclonic jerk, which made it seem like they were falling before they actually fell, waking them up and making sure that they had a well-balanced and safe position before they resumed their arboreal sleep. If Sagan was right, then myoclonic jerks are somnambulistic behavioral fossils, just like the ninety-minute sleep cycle is a remnant of our earlier night watches.

The interesting thing about myoclonic jerks is that they are infective, or perhaps more accurately, are conductive—they can travel from one sleeping person to another if two people are in close contact while they are falling asleep. Many married couples report the phenomenon of being unable to tell in whose body the myoclonic jerk originated, the transmission of the jerk from body to body is that instantaneous.

When atonia doesn't work, things get interesting, and sometimes dangerous. Without the safety net of sleep paralysis, dreamers can act out their dreams physically. Sleep talking and sleepwalking (also known as somnambulism) are common results of this failure. Fifteen percent of children under the age of twelve sleepwalk at least once, though sometimes somnambulism continues into adulthood. Rare individuals have a frighteningly active form of somnambulistic disorder that sometimes leads to violence. One patient of a sleep disorder clinic in the United States, an ex—football player, awakens his wife some nights when he tackles the dresser in their bedroom. But sometimes the sleeper's violence is frighteningly focused, and murder can result. There have been several famous homicide acquittals based on the defense that the perpetrator was asleep at the time of the murder. The first such acquittal occurred in 1846, when Albert Tirrell was found innocent of the murder of Marie Ann Bickford, a prostitute that he had left his wife and family for. Apparently she refused to give up her profession and one night, in his sleep, Albert slit her throat and set fire to the brothel in which she was working. An amusing, though possibly apocryphal, Victorian anecdote from the writings of Augustus Hare neatly captures the half-sinister, half-ludicrous nature of somnambulism: "A lady was awoke in the night with the disagreeable sense of not being alone in the room, and soon felt a thud upon her bed. There was no doubt that someone was moving to and fro in the room, and that hands were constantly moving over her bed. She was so dreadfully frightened, that at last she fainted. When she came to herself, it was broad daylight, and she found that the butler had walked in his sleep and laid the table for fourteen upon her bed."

A Visit to the Sleep Laboratory

While researching slumber I decided to see what goes on in a sleep laboratory firsthand. I got a referral from my family doctor and I booked a night at the well-known St. Michael's Hospital Sleep Laboratory in Toronto. My appointment was for a Tuesday night in early October, and a few weeks before my visit I received a letter with a series of instructions and a map showing me how to get to the clinic. I'd be able to bring my own pillow, pajamas, and overnight things, as well as reading material to keep me occupied. They recommended that I check in at 7:45 P.M., but because I was taking a night-school course in Italian, I phoned and got permission to arrive later, at 9 P.M.

The night I went to the sleep clinic was clear and cool. I parked in a multilevel garage across the street from the clinic that was practically empty and I felt a little self-conscious taking my pillow and overnight bag out of the trunk. I had a feeling of anticipation that I hadn't had since I was a kid, similar to the ambivalence I felt on the eve of a sleepover at the home of an aunt I didn't know that well. The woman at the admissions desk was middle-aged with bleached blond hair and long fingernails with decals on them. She looked up, glanced at my pillow, and said, "Sleep clinic?" "Yes," I answered, holding up my pillow. Her fingernails clicked across the keys as she entered my personal data and printed out an embossed, plastic card that I needed to present at the laboratory.

After a short elevator ride I entered the laboratory itself. A clean, modern foyer opened to a curtained prep room on my left, the laboratory and patient-observation room to my right, and then, down a hall stuffed with expensive and complex-looking computer equipment, a series of six bedrooms. A technical assistant showed me into one of them, my designated sleeping room, and told me to get into my pajamas and come back out when I was ready. The room was quite adequate, a large window opening to a view of the city at night, a spartan en suite bathroom, a bed, and a small desk. Beside the bed was a trolley with drawers and an electronic device on top, and on the ceiling above the bed I noticed a small video camera pointing down at where I would be sleeping.

My technician knocked discreetly after I finished putting on my pajamas and then led me to the prep room. He explained everything as he went along and I peppered him with questions. I asked him what he was called professionally and he replied that he was a "polysonographologist." As we talked, he began to tape electrodes onto my calves. To monitor restless-leg syndrome, he said. Then he taped electrodes to my temples (to detect eye movement when I dreamed), on my scalp (to measure my brain waves and determine what stage of sleep I was in), and on my chin (in case I was a tooth grinder).

The electrodes, with their fine wires, seemed pretty delicate. "I sleep on my stomach," I told him. "That's okay," he said, "just don't do a three-hundred-and-sixty-degree roll or you'll rip out the connections." "I'll try and remember," I replied. After he was through, I was a mass of wires and electrodes. He draped a connector and plug around my neck and steered me back to my room. "I'll be in just before lights-out to attach your chest and abdomen bands, as well as a nasal canula," he said. I wasn't quite prepared for that. "What's that exactly?" I asked. He described the clear, flexible plastic breathing tubes that fit just under your nose when you need oxygen in the hospital. I conjured up the image from some TV show I'd seen and I nodded.

I went back to my room and put the questionnaires the technician had given me on the small desk. Beside it was a window with a view of one wing of the hospital and the lights of the city beyond. I remembered some of the sleepless nights in the hospital when I had my appendix removed in my early thirties, the quiet wards, the anxious coziness. I wondered who was looking out of some of those windows in the hospital, looking at the night with despair or hope. I sat down and worked at my desk filling out some questionnaires about my sleeping habits. They contained questions like: "How disturbed is your bed partner by your snoring? 1) not at all; 2) mildly, 3) moderately; 4) severely," and "Has anyone ever told you that you stop breathing when you sleep? Yes or No."

I completed the questionnaires and did some reading until it was time for lights-out. My technician knocked again, very quietly, and I let him in. He adeptly and quickly taped the canula under my nose, though the two nostril extensions were quite tickly. "That ticklish feeling will go away after a few minutes," he said. He also taped a tiny, glowing red light to my right ring finger. "We call that the E.T. light. It measures blood flow and oxygenation," he explained. After putting two elasticized bands on me, one around my abdomen, the other around my chest (to measure breathing), he plugged in all the leads to the console beside the night table and told me to turn out the lights within half an hour. "What if I have to pee in the middle of the night?" I asked him. "Just use that buzzer there beside the bed, and I'll come in and unplug you," he said. I hoped I wouldn't have to.

When the lights were out, I put my right hand up in the air and admired my glowing red fingertip. I tried to sleep on my back at first, but eventually I turned to my left side and then, finally, to my stomach. The nasal canula tickled every time I moved and the plugs for my abdominal band were uncomfortable, not to mention all the wires trailing over my body. I fell asleep, fitfully, for a couple of hours until around 4 A.M., when I decided that I had to go to the bathroom.

I buzzed and my technician came in, undid my connections, and after I was finished, came back in and replugged the leads. I didn't sleep that well, and I woke up several times. I thought of them looking at me through the camera and realized that, even if I didn't open my eyes or move at all, they still knew I was awake from my brain waves. They were inside my head. It was sort of spooky and vaguely invasive— thank goodness this had all been voluntary because the scenarios for various exotic sleep tortures began to run through my mind. I looked forward to 6 A.M., when I could leave.

In the morning my technician knocked promptly at six. He knew I was awake. After he had removed the electrodes and I had dressed, he gave me a tour of the control room where they monitored the patients. There was a row of six television screens and in a couple of them I could see that two women were still sleeping, the wires trailing out from under their sheets just as mine had. There were also several computer screens that the technicians could use to switch from patient to patient, monitoring their brain waves, their heart rates, breathing, oxygen levels, and leg movements. I asked my technician how they coordinated all this data. "Everything is dumped into the central memory," he replied, the videos, the encephalograph, every- thing. "That must be some operating system," I said. "Yes, it is," he answered, "it's registered as Mr. Sandman."

A few weeks later, on a late November afternoon, I went back to meet with Dr. Hanly, the head of the sleep laboratory, to review my charts. A receptionist pointed me toward a small examination room, and after I sat down and looked around the room, I realized I was a little nervous. What if, by some bizarre coincidence, I had severe sleep apnea or unusually loud snoring! I didn't have to wait long before Dr. Hanly entered the room. He was a tall, precise man in his late forties with graying hair and a slightly distanced demeanor. He had an intern with him and asked if I minded having the intern there and I told him I didn't.

Dr. Hanly listened to my breathing with a stethoscope and measured my blood pressure, both of which were, apparently, normal. Then he asked me to stand up and open my mouth wide. He looked into my throat. He got the intern to look also. "See, perfectly normal," he remarked to the intern, to my private relief. Then he asked me to sit down, and we went over the data from the sleep clinic. He asked me how long I thought I'd slept and I replied, "One, maybe two hours." I was surprised when he said, gesturing at his sheaf of computer printouts and charts, that I had slept for three and a half hours.

He then told me that I had exhibited some symptoms of minor sleep apnea. I was shocked. Like most sleepers, I'm not even aware that I snore, even though I've been told that I do. I asked him if my apnea was serious and he reassured me that my oxygen levels never dropped below normal (the image of the glowing red E.T. finger came back to me) and that I wouldn't need any treatment at all, not even the jaw-adjusting prosthetic device that he sometimes recommends for minor apnea.

After that he went over the rest of the charts, which were completely normal except for one spike that he pointed out with a pen. "See that?" he asked. What now? I thought. "This is a transient peak of sixty decibels," he said, "a little high, but otherwise your snoring is well below the antisocial range." (According to the *Guinness Book of Records*, Mel Switzer, an Englishman from southern Britain, has the loudest snore ever recorded. It has been measured at ninety-two decibels, equivalent to a big motorcycle starting up from a distance of two feet. His wife, conveniently, is deaf in one ear.)

I thanked him for his time and, after shaking hands once again, I left the room and went to the elevator. I thought about the unexpected diagnosis, which seemed like a bad thing, and the fact that it didn't require any treatment, which seemed like a good thing, with the requisite ambivalence. From the research that I'd done for this book, I knew that apnea could be serious, that it could cause exhaustion and heart disease. Yet Dr. Hanly didn't seem too worried, so I decided during the elevator ride that I could wear the badge of my diagnosis with some bravura, perhaps use it as a topic of conversation at a dinner party, as a symptomatic weapon in some hypothetical battle with others who had more exotic disorders.

By the time I got back home it was dark. Walking up the path from my garage to my house, I recalled the squiggles on the graph sheets that represented my dreams, and I realized that technology still has no way of actually looking into our heads at night to see our dreams. Those electroencephalograph lines correlated to events in my dreams but were only the electronic shorthand of my experiences. We cannot record our dreams and replay them the next day; the only way we can revisit our dreams is through memory. It is all in our head.

Dreaming

Being primarily creatures of the day, like other diurnal animals, we spend most of the night asleep. But we are complex beings, and most of our complexity lies within our skulls. Our brains are ceaselessly active when we are awake, tracking our visual environment, reacting emotionally to events and memories, and ruminating unconsciously about our own lives and the lives of those around us. Our job stress and preoccupations mount up throughout the day, and if it were not for sleep and dreams, we would become psychotic or unbalanced. Torturers have used this fact to force confessions out of prisoners, and sleep deprivation has been used for centuries to crack the most tight-lipped spies.

Our brains are like complex, nervous sponges that soak up the outside world during the day, but at night when we close our eyes, the sponge is squeezed by sleep and the day leaks back out. Night is like a film-editing room of our unconscious minds where the film of the day is played back. As James Boswell wrote, "In sleep the doors of the mind are shut, and thoughts come jumping in at the windows. They tumble headlong, and therefore are so disorderly and strange." But dreams are more than just the random replay of recent memories; they follow a deeper flow and seem to have their own logic. Ralph Waldo Emerson once observed, "Dreams have a poetic integrity and truth. This limbo and dust-hole of thought is presided over by a certain reason, too. Their extravagance from nature is yet within a higher nature."

Ultimately dreams are images and signs that surface after a shape-shifting journey through the hidden contours of the mind. They can tell us more about our selves than we think we already know, though their interpretation is as slippery as their veiled narratives. Michel de Montaigne, a keen observer of dreams very much ahead of his time (the sixteenth century), wrote, "I hold that it is true that dreams are faithful interpreters of our drives; but there is an art to sorting and understanding them."

8

EMPIRE OF DREAMS—12 A.M.

*Is not the midnight like Central Africa to most of us? Are we not tempted to
explore it,—to penetrate to the shores of its Lake Tchad, and discover the source
of its Nile, perchance the Mountains of the Moon? Who knows what fertility
and beauty, moral and natural, are to be found? In the Mountains of the
Moon, in the Central Africa of the night, there is where all Niles have their
hidden heads.*

HENRY DAVID THOREAU

IT IS NOW MIDNIGHT. The sun is on the opposite side of the
planet, directly beneath our feet, and it casts the earth's shadow
directly above us like a dark, invisible searchlight into the night sky.
Twelve A.M. is also the technical beginning of morning—every day
begins in the night and ends in the night. But even though our ideal
night is half over, midnight marks the beginning of another kind of
night, the deep night, the "small hours" when most people are
dreaming in their beds. If any part of night is an unexplored
continent, as was the heart of Africa during Thoreau's time, then
it is deep night.

Midnight is the hour of romance and magic, as Shakespeare
observed in *A Midsummer Night's Dream*: "The iron tongue of midnight hath told twelve / Lovers, to bed; 'tis almost fairy time."
Although it's also sometimes referred to as "the witching hour,"
midnight has definitely seen as much romance as it has broomsticks.
Lovers wait until the "Midnight Hour," as Wilson Pickett used to
sing, because that's when their love comes "tumblin' down." Romantic couples rendezvous " 'round midnight" in the same cool jazz

clubs Thelonious Monk used to play. Night's ultimate hour has inspired other tunes as well: "Midnight Train to Georgia," "Midnight at the Oasis," and "Midnight in Moscow" come to mind, but there are other cultural associations with the witching hour that are more practical. We go to midnight-madness sales and munch midnight snacks by the light from an open refrigerator. The pigment called midnight blue is similar to cobalt blue, and many painters burn the midnight oil as they work late at night. In the land of the midnight sun, sleepers have to draw their shades at night. Midnight is also a magic moment when occult rituals are held, though for Cinderella it was curfew time. A midnight mass is held on Christmas Eve and Paul Revere rode out at midnight.

For those nighthawks who work best in the evening, midnight marks the beginning of uninterrupted work. Cortisol, the stress hormone, is at its lowest levels just before midnight, and those who remain awake enjoy a second wind. Poet Anna Barbauld wrote about the peculiar energy of midnight at the turn of the eighteenth-century: "This dead of midnight is the noon of thought, / And Wisdom mounts her zenith with the stars." Like many writers, Milton also seemed to do his best work at night. For Milton it was a badge of honor that he kept working late into the night: "Let my lamp, at midnight hour, / Be seen in some high lonely tower." American poet Charles Olson was also a night owl, famous for his nocturnal lectures at Black Mountain College in North Carolina. Sometimes his seminars didn't start until 11 P.M. When he was composing poetry, he drank scotch and often wrote until dawn, describing the literary night shift as part of a zany ecology: "by night only crazy things like the full moon and the whippoorwill and us, are busy."

Dreams

Breathe slumbrous music round me, sweet and slow,
To honied phrases set!
Into the land of dreams I long to go.
Bid me forget!

MARY E. COLERIDGE

We spend the majority of night sleeping, and even though new research has given us a glimpse into what goes on in our heads at night, the realm of sleep and dreams remains a largely unexplored territory. It is only recently that we have gathered any concrete knowledge about the activity of the brain during slumber. Until Dr. Nathaniel Kleitman's groundbreaking studies in the early fifties, sleep was regarded as a resting period for the brain and it was generally presumed that its activity slowed to a crawl during slumber. Now we know that the dreaming brain is very active.

We have dreamed as long as we've been human, and probably before that. Dreams have inspired emperors and artists. They have changed military history, provided breakthroughs for scientists (the discovery of the structure of the benzene molecule was inspired by Friedrich August Kekulé von Stradonitz's 1890 dream of an oroborus), and have carried communications from the gods. The power of dreams is uncontested and their influence has echoed throughout our history.

The often bizarre logic of dreams is something all of us have probably encountered when reawakened at the edge of sleep, that moment just as we are beginning to dream when something interrupts. It has happened to me several times and there has always been some crazy connection between the last thing I was thinking about before I went to sleep and the images in my dream. It appears that dreams are indeed a continuation of our train of thought, our interior monologue, but it's as if the tracks that our associations run on during the day disappear during sleep and our train of thought is derailed. This is partly because as we fall asleep our network of mental connections begins to respond to internal stimuli: hormone-initiated emotions, leftover moods, indigestion, and bladder capacity. It's as if our chain of images was still in a semiconscious mode, but without an external world to influence them, they continue on their way without guidance or direction.

Another way of understanding dreams is to consider how our brains work. When we are awake, our brains are constantly responding to the information from the outside world that floods into our heads. The very reason that our brains exist is to process the sensory and social data that come from the external world. At night, they are cut off from this sensory flow; our brains are like fish out of water. That is why our sense-starved brains create their own worlds when we are asleep. The movie cameras of our eyes are shut off and our minds become projectors, throwing their own films upon the blank screen that arises when sound, sight, and even bodily awareness are taken away by sleep. Federico Fellini once commented on the resemblance between movies and dreams in an interview in *Rolling Stone*: "Talking about dreams is like talking about movies, since the cinema uses the language of dreams; years can pass in a second and you can hop from one place to another. It's a language made of image. And in the real cinema, every object and every light means something, as in a dream." The only amendment I'd make to Fellini's observation is that while dreams are indeed like watching a movie, it is a movie that you are unconsciously producing—you cast the roles, you film the scenes, you edit, and then you project the film at the very same time as you are watching it! Furthermore, it is a movie in which the twists and turns and very ending are completely dependent on your emotional reaction to it. Its trailers are your daydreams and its theme consists of all your hopes, lusts, fears, and terrors.

Interpreting Dreams

The earliest recorded dreams date back almost four thousand years ago to ancient Mesopotamia, the spawning grounds of civilization and the first written languages. These first oneiric diaries were discovered during a dig in Iraq in the mid nineteenth century when archaeologists unearthed the Royal Assyrian Libraries of Nineveh in the form of hundreds of clay tablets with inscriptions on them. This writing, which we now know as cuneiform script, defied translation for decades. In fact, it took another forty years before George Smith of the British Museum cracked the code. But even more fascinating was Smith's discovery that contained within these hundreds of clay tablets was an amazingly intact piece of early literature, the work of a Babylonian poet who lived more than three thousand and seven hundred years ago. The writing turned out to be an epic poem about a legendary king called Gilgamesh, and the whole work is now known by his name. It is here, in the chronicles of Gilgamesh, where the first unequivocal accounts of dreams were found.

According to the epic, Gilgamesh had a series of recurring dreams that he reported to his goddess-mother, Ninsun. Ninsun interpreted his dreams (the first recorded dream interpretations) as prophecies that had direct applications to Gilgamesh's administration of his empire. This prognosticative notion of dreams is a thread that will recur throughout dream interpretation from Mesopotamia right up to the nineteenth century. The Mesopotamian system of dream interpretation was disseminated throughout the region with the spread of the great city-states, and found its way first to Egypt and then to India.

The Indian history of dream interpretation is almost as old as the Mesopotamian and Egyptian histories, though the Indian system is so unique that ethnoarchaeologists believe that much of it was indigenous. A section of the Vedas (The Sacred Books of Wisdom), which

were written between 1500 and 1000 B.C., deals with dream interpretation. According to Vedic lore, dreams that included violence
were interpreted as good omens for success and happiness, though
only if the dreamer was an active participant. In some respects this is
sound psychological advice, because people who try to control
dangerous situations in their dreams are probably more disposed
to positive action in their waking lives. The Upanishads, written
later, between 900 and 500 B.C., continued the tradition of the Vedas.
According to the Upanishads, dreams are the manifestations of the
dreamer's desires, though they also claimed that the soul left the body
while it was asleep (as did the Chinese). The writers of the Upanishads
believed that if a dreamer was awakened too abruptly, his or her soul
would be caught outside the body and the sleeper would die. (This
notion continues today in the superstition that a dreamer who is
awakened too quickly might die of shock.)

The Hebrew Bible, written at about the same time as the
Upanishads, had a much more religious explanation for dreams.
Its writers believed that dreams were communications from God
and many Old Testament figures, from Solomon and Nebuchadnezzar
up to Joseph, were visited in their dreams either by God or prophets.
The advice they received was often applied to a crucial decision and
helped guide their actions:

> "In a dream, in a vision of the night, when deep sleep falls on mortals,
> while they slumber in their beds, then he opens their ears, and terrifies
> them with warnings, that he may turn them aside from their deeds,
> and keep them from pride, to spare their souls from the Pit, their lives
> from traversing the River."
>
> JOB 33:15–18

Jacob's visionary dream of a ladder reaching from the earth to
heaven is perhaps the most famous of these ancient Hebrew chroni-

cles. It is worth quoting because the dream vision conjoins night, holy
ground, and an Old Testament mood of awe, revelation, and spooky
magic.

> Jacob left Beer-Sheba, and set out for Haran. He came upon a certain
> place and stopped there for the night, for the sun had set. Taking one
> of the stones of that place, he put it under his head and lay down in
> that place. He had a dream, a stairway was set on the ground and
> its top reached to the sky, and angels of God were going up and down
> on it.
>
> GENESIS 28: 10–22

Viewing dreams as messages from God reflects the regional beliefs of
the eastern Mediterranean during this period of history. The dreams of
Egyptian royalty were also said to be divinatory, and Egyptian gods
would often appear in them. Serapis, the Egyptian god of dreams, had
temples dedicated to him, and within them were special dream-
incubation rooms. It was thought that sleepers dreaming within an
incubation room would receive wisdom and advice from Serapis.
Prospective dreamers would undergo a period of purification prior to
visiting the incubation chamber during which they fasted and prayed,
a process that was said to ensure revelatory dreams. (The intimate
relationship between fasting and dreams was paralleled in the Native
American culture of North America in precontact times. Prior to
"vision quests," Algonquian initiates had to fast several days before
they had their "power dream," in which their unique totem animal
was revealed to them. The totem would act as their guardian, and its
powers would be transferred to the dreamer.)

Curiously enough, the Chinese, like the Egyptians, also practiced
incubation in dream temples. They believed that the spiritual soul,
which they called Hun, left the body while the dreamer lay asleep and
that it communicated with the land of the dead. The Chinese dream-

incubation temples were incorporated in a pragmatic way with municipal and governmental affairs. Visiting and local officials, including judges, were obliged to use the incubation rooms for guidance and wisdom.

It is with the Greeks that the first glimmers of a modern understanding of dreams appears. Antiphon of Athens wrote the earliest-known Greek text on dreams. It reflects the mainstream beliefs of the time (around 500 B.C.), namely that the soul of the sleeper left the body and that dreams were a record of this journey. Greeks also thought that the gods visited dreamers while they slept. They did this by entering, and leaving, through a keyhole. It is this belief that must have underlain the Greek myth about Zeus, who visited one of his many paramours by pouring himself through a keyhole in the form of a shower of gold. The Greeks also practiced dream incubation, a technique they probably inherited from the Egyptians. Specialized sanctuaries dedicated to the Greek healer Aesculapius sprang up around Athens in this period. It was thought that Aesculapius, believed by many to be the son of Apollo, would visit incubation chambers and cure the dreaming sleepers there. A priest would interpret the dreams of patients as descriptions of remedies, prescribed by the gods, for whatever ailment brought the person to the incubation chambers.

Modern dream interpretation starts, as does so much else, in the fifth century B.C. with the Greek philosophers. It was Heraclitus who first suggested that dreams might be products of the sleeper's mind rather than messages from the gods, and Hippocrates (469–399 B.C., a contemporary of Socrates), who was regarded as the "father of medicine," wrote a strikingly modern-sounding, psychological passage where he explained dreams in a single sentence: "By day the soul receives images and by night it produces them." This is in complete agreement with modern psychological theories on dreaming, though it was Aristotle (384–322 B.C.) who began the first rational study of dreams. In his essay "De Divinatione Per Somnum," he debunked the popular belief in the divinatory power of dreams by writing, "most so-called prophetic dreams are to be classed as mere coincidences, especially all such as are extravagant." Later in the same essay he even anticipates modern psychological interpretations, claiming that "the most skillful interpreter of dreams is he who has the faculty of absorbing resemblances. I mean that dream presentations are analogous to the forms reflected in water." Or, in other words, dream images are both symbolic and elusive. In his *Parva Naturalia*, Aristotle anticipated the Freudian concept of "day residue" where dreams formed around the events of the preceding day: "dreams are in fact believed to be a recollection of the day's events."

The Romans continued the Greek philosophical tradition, though they brought to it their own special talent for collecting and anthologizing. As a result, one of the largest historical collections of dream interpretations was the *Oneirocriticon* ("interpretation of dreams"), written by Roman author Artemidorus of Dalis in A.D. 150. The five-volume *Oneirocriticon* was the single biggest compilation of dream interpretation that had ever been put together. Artemidorus anthologized all the standard folk interpretations of his day from as far afield as North Africa, Greece, and the Middle East, as well as introduced a few of his own insights, including the notion that dreams were unique to the dreamer and that dream symbolism was contingent on the dreamer's social class, health, and occupation. Otherwise, most of Artemidorus' dream interpretations focused on predictions of personal financial gain or loss.

Interest in dreams began to wane with the rise of Christianity, though the Christian philosopher Augustine did speculate about dreams in several of his books. He believed that dreams were a sort of clue to the interchange between the trinity of body, mind, and heaven. In a letter he wrote to Nebridius in A.D. 389, he postulated that there might be purely imaginary beings: "you will see that it is possible for us both in dreams and in waking thoughts to conceive the images of bodily forms which we have never seen." Unfortunately, at least for the evolution of dream interpretation, the Dark Ages were appropriately named. After Augustine little was written about dreams and during the European Middle Ages dream interpretation fell further into disrepute as the Christian church was deeply suspicious of the practice. If dreams were studied at all, they were viewed only in the context of religion. In fact Martin Luther, the patriarch of Protestantism, believed that dreams were authored by the devil and that sin was "the confederate and father of foul dreams." It wasn't until the nineteenth century that the Western world caught up with dream interpretation, and then it did so with a vengeance.

The Royal Road to the Unconscious: Freud's Interpretation of Dreams

This is not a topic for discussion at a scientific meeting; it is a matter for the police.
PROFESSOR WILHELM WEYGRANDT ON FREUD'S THEORIES
AT THE CONGRESS OF GERMAN NEUROLOGISTS AND
PSYCHIATRISTS, HAMBURG, 1910

In the summer of 1895 Sigmund Freud and his wife rented a villa on the outskirts of Vienna. Freud had been working hard over the winter and spring and this sojourn was exactly what he needed to reacquaint himself with his young family. His psychoanalytic practice was burgeoning and his reputation was growing not just in Austria but also in the rest of Europe. The contemplative time that this holiday provided was essential for Freud: his research into the unconscious mind had produced a puzzling series of possibilities, and if he was to decode the organization of human consciousness, he needed to retreat some distance to get an overview.

The villa that the Freuds rented, called Schloss Belle Vue, sat atop a hill and commanded a good view. After dinner Freud often smoked a cigar outside in the twilight and ruminated alone. When he finished his cigar, he'd retire to his study to review his analysis notes and read. The restorative calm of the holiday proved to be very fertile, for on the night of July 24, several weeks after arriving, Freud had a dream that changed the course of modern psychology. The dream itself was not that unusual; it involved a patient that he was currently analyzing called "Irma." In the dream she appeared pale and sickly and Freud decided that her condition was due to malpractice by a colleague of his. When he woke up, he realized that his own guilt about her condition had been transferred to his colleague, and that her sickly condition was a symbol of his remorse over a misdiagnosis. The dream had neatly solved his wish to wash his hands of responsibility for her illness by displacing the blame on his colleague's shoulders—in short, it was an excellent example of wish fulfillment, which, in turn, became one of the cornerstones of his dream interpretation technique.

When Freud discovered that dreams were predicated on a deferral of latent content into manifest content—that is, that almost everything we visualize in our dreams refers back to an emotion or drive—he had one of those eureka moments that all scientists hope for. Freud began to write his seminal work, *The Interpretation of Dreams*, almost immediately. Years later he stated that his Irma dream was so momentous that a historical plaque should be affixed to Belle Vue reading; "In this house on July 24th, 1895, the secret of dreams was revealed to Dr. Sigmund Freud."

Before Freud's revelations, Victorians could console themselves that even if they weren't at the center of the universe—Galileo had accomplished this displacement—at least they were masters of their own minds. But Freud added yet another installment in a series of traumatic blows to the primacy of humans with his claims that not only were we not masters of our own minds, we were also merely the

puppets of our unconscious minds, and that we hardly knew ourselves. Freud was thrust into the center of a storm of controversy.

In *The Interpretation of Dreams* Freud described dreams as consisting
of several central components: wish fulfillment, dream work, displacement, day residue, and condensation. The first of these, wish
fulfillment, is probably Freud's most basic observation about dreams,
and one that we all recognize—many of our dreams are nocturnal
fantasies about things we'd like to happen. The second component,
his concept of dream work, is a little more complex. Freud called
dreams "picture puzzles" that had to be interpreted, that is, the
content of the dream was symbolic of the dream's "intention." This
translation of an intention into a symbol by the dream is what Freud
called the dream work. He then designated the term *displacement* to
refer to what was achieved by the dream work; the dream was always
"centered elsewhere," as Freud wrote, as if the dream were actively
trying to throw us off the scent. As for Freud's designation of day
residue, it's something we've all experienced in our dreams, namely a
reference to an actual event that occurred in our waking lives, often on
the day prior to the dream. Finally, condensation refers to the
tendency of dreams to compress several different things into one
object, such as a dream of a single person who has the characteristics of
two or three other people whom the dreamer knows.

Shortly after its publication *The Interpretation of Dreams* found its
way to the prestigious Burgholzli Clinic in Zurich, Switzerland,
where it changed the life of a twenty-five-year-old psychologist who
worked there. His name was Carl Jung and he immediately wrote to
Freud, initiating a friendship and eventually an apprenticeship that
would last until their bitter estrangement in 1912. Jung would go on
to promulgate his own, somewhat different interpretations of dreams
and the unconscious mind, but before their final scission ended their
collaboration, Jung and Freud had taken a trans-Atlantic cruise to
America, where they addressed a congress of American psychologists.

One night, while aboard the steamer, Jung had a dream that became the cornerstone for his own unique method of dream interpretation. In this dream he was exploring an old house with many floors. As he moved downward through the house, he noticed that each floor was furnished with furniture and objects from increasingly older periods of history. In the basement he found a trapdoor that led to a prehistoric tomb filled with human skulls and ancient pottery, much like the Egyptian tombs that the archaeologists of his day were discovering.

It is from this dream that Jung derived his theory of the collective unconscious, namely, that we all have access, through our unconsciousness, to the collective myths and archetypes of all human beings. This insight became the cornerstone of his theories of the unconscious and his theory of the archetype, for which he became famous. Jung was more concerned with what the dream had to say for itself than a "top-down" interpretation of the dream, and he granted the unconscious much more autonomy than Freud ever had. In his book *Dreams*, Jung claimed that every dream had four stages that proceeded in a fixed order. They were: statement of place, development of plot, culmination, and finally, a solution or result. The statement of place, of course, was the location of the dream and the development of plot was what led up to the culmination (the central scene) of the dream. The solution was what Jung saw as the end point of the dream work—that is, a synthesis of what the dream was trying to communicate to the dreamer: its resolution.

Jung's theories strayed from Freud's in that Jung felt that dreams were part of a psychic immune system and that they had a strong orthopsychic, healing potential. In his book *The Psychology of Dreams*, Jung wrote: "Just as the body reacts in a purposeful way to wounds or infections or an abnormal way of living, so the psychic functions react to unnatural or injurious disturbances with appropriate means of defence. One of these purposeful reactions is the dream, in which the unconscious material constellated round a given conscious position is presented to the conscious mind in symbolic form."

Dream Analysis

Your vision will become clear only when you can look into your own heart . . .
Who looks outside, dreams; who looks inside, awakes.

CARL JUNG

Jung's legacy has been the practice of dream analysis, a type of psychotherapy sometimes blended with his own brand of psycho-analysis, called analytical psychology. I underwent dream therapy myself a number of years ago and found the experience extraordinarily revealing. The technique is very simple, though the results are powerful. Every morning I would write down the dreams that I remembered on a notepad beside my bed, and twice a week I would discuss my dreams with my dream analyst.

The amazing thing about this process was the steady increase in my ability to remember my dreams in all their complexity and details over the course of the analysis. The paragraph or two of description I jotted down each morning when I first started analysis soon bur-geoned to several pages of notes, which sometimes required an hour to write down, so complete and exhaustive did my dream memory become. It was as if the process of dream analysis gave me a key to my dreams, a key that grew more elaborate and opened more and more rooms as the year went on. I began to recognize themes and under-currents that linked dreams that had occurred months apart into installments of a bigger narrative.

I still have the sheaf of all my dreams, which I transferred into type from my notes, along with drawings of pertinent scenes and objects. And even now, years after I stopped my dream analysis, I go back to them every now and then and gain some new insight into myself that completely escaped me at the time. It is as if my old dreams are becoming more and more transparent to me as I become more objective with distance, their former opacity turning into a clear medium within which I can see what I was trying to tell myself then. (Curiously, these insights into my old dream patterns do not always translate into a deeper understanding of my present dreams.) The interesting thing about reading my old dreams, even now, is how the mood of these dreams, their feeling or "affect," as the psychoanalysts call it, comes back to me as freshly and as mysteriously as if I had just dreamed them. There is something magical and elusive about dreams that continues to defy analysis. As Charles Baudelaire wrote in *Artificial Paradise* in 1860: "To dream magnificently is not a gift given to all men, and even for those who possess it, it runs a strong risk of being progressively diminished by the ever-growing dissipation of modern life and by the restlessness engendered by material progress. The ability to dream is a divine and mysterious ability; because it is through dreams that man communicates with the shadowy world which surrounds him. But this power needs solitude to develop freely; the more one concentrates, the more one is likely to dream fully, deeply."

After Jung and Freud

> *Sleep hath its own world,*
> *And a wide realm of wild reality.*
> *And dreams in their development have breath,*
> *And tears, and tortures, and the touch of joy.*
> GEORGE GORDON NOEL BYRON

Perhaps the most significant contribution to dream interpretation after Carl Jung was made by Calvin Hall, a psychologist who developed a cognitive theory of dream interpretation in the early 1950s. Hall regarded dreams as maps that reveal paths to the dreamer, and curiously, resurrecting the predictive view of dreams, that these paths were like instructions showing us how to anticipate problems in our lives.

Hall believed that dreams embodied the dreamers' private relations to themselves, their family, and their friends. He also developed a series of character types that summarized how individual dreamers dealt with their dream worlds. These categories included assertive, weak, domineering, unloved, and hostile dreamers. His major contribution to dream interpretation was his notion of dream metaphors, which, similar to the work of French semiotician and psychoanalyst Jacques Lacan, asserted that dreams were often based on language and used puns, metaphors, and visual plays on words. An anecdote from the life of famous neurologist J. Z. Young neatly illustrates the action of metaphors in dreams. In his book *Programs of the Brain* he related that "when I was about to retire from University College I dreamed with dread of having to give a lecture in Helsinki. 'Why in Finland?' I asked my wife. She saw the symbolism at once, 'It's the finish of life!' "

Hall's work also confirmed Jung's idea of a collective archetype: his empirical research of dreams from all over the world showed an extraordinary similarity. As well, his research also revealed a direct link between conscious concerns and dream content and showed that certain individual's dreams demonstrated a remarkable continuity of themes over many years. But it turns out that there are individuals with even more unique capabilities, dreamers who have the ability to control their dreams.

Lucid Dreams

Saddle your dreams afore you ride 'em.
MARY WEBB

Lucid dreaming, in which the dreamer is aware that he or she is asleep, is a much more recent and experimental area of dream research, though Henry Thoreau seemed familiar enough with it when he wrote, "Our truest life is when we are in dreams awake." The concept of lucid dreaming gained widespread recognition in 1867 with the publication of *Dreams and How to Guide Them*, by Marquis Marie Jean Léon Hervey de Saint-Denys. Marquis Hervey de Saint-Denys was a famous lucid dreamer and could not only control his dreams but also claimed to be able to move fluidly between his waking world and his dream world. He was also able to continue interrupted dreams at will and remember to do things in dreams that he thought of while awake. He recounts how one night he dreamed that

> on coming out of a theater, I got into a hackney carriage, which moved off. I woke up almost immediately, without remembering this insignificant vision. I looked at the time on my watch; I picked up a lighter that I had knocked over; and after having been *completely* awake for ten or fifteen minutes, I fell asleep again. It was here that the strange part of the dream began. I dreamed that I woke up in the carriage, which I remembered perfectly well having entered to go home (in the dream). I had the impression that I had dozed off for about a quarter of an hour, but without remembering what thoughts had passed through my mind during that time. Thereupon I reflected that a large part of the journey must already be over, and I looked out of the window to see what street we were in, having thus interpreted as time asleep the very time when I had stopped sleeping.

Night for Hervey de Saint-Denys was home to a fantastic world as lucid and complex as the world of his waking life, and it took science almost a century to catch up to him.

Our modern concept of lucid dreaming was first established in 1913 by Dutch psychiatrist Frederik Van Eeden, and it has been a marginal, if not contentious area of dream research ever since. The Lucidity Institute in Palo Alto, California, is currently the preeminent lucid-dreaming research institute. It was founded by Dr. Stephen LaBerge, a psychologist from the Stanford University Sleep Center, and the main thrust of its research at this point is to prove the therapeutic value of lucid dreaming. Although a difficult technique to master, lucid dreaming can enhance the life of the dreamer in a positive way, and, according to Dr. LaBerge, the dream world can become a nurturing environment for personal empowerment. The future of dreams lies perhaps in our ability to control them, both as tools of self-knowledge and enhancement, and perhaps recreationally, as a sort of virtual reality in which we have omniscient powers. But dreams still have the upper hand, and they seem especially powerful when they terrify us.

Nightmares

The fourth edition of *The Psychiatric Dictionary* defines nightmares as "a fright reaction during sleep," but for anyone who has experienced a severe nightmare, it is more like falling through a trapdoor into a pit of horror and psychic nausea. If we are indeed the authors of our dreams, then the perennial question that nightmares evoke is: where did *that* come from? Nightmares are often quite vivid and are usually characterized by feelings of helplessness, suffocation, and a sense of panic. Not only that, the sense of dread that accompanies nightmares can be reawakened the following day if the nightmare is recalled. Nightmares usually strike within the first hour or two of sleep and almost always cause their victim to wake up, sometimes bathed in sweat. Common nightmare themes among adults include falling, being chased, and being kidnapped. Natural disasters such as tidal waves, tornadoes, floods, and earthquakes occur frequently. Of these calamities, the ones involving water are the most significant, according to Jungian dream analysis, because they deal with the unconscious itself, although nightmares about drowning are more concerned with the fear of suffocation. Technological disasters, including nuclear war, chemical or biological attacks, explosions, and fires are common nightmare material, as is being a passenger in an airplane or automobile that is out of control and about to crash. Interpersonal nightmares often involve themes of betrayal, abandonment, and humiliation, as well as incidents of violent attacks or attacking and injuring or murdering someone else. Spooky nightmares tend to include ghosts of dead relatives or acquaintances returning in a horrible form. Being paralyzed or unable to move or scream for help is common, as is being fatally ill with an incurable disease or having various parts of your body fall off or detach in a gruesome manner.

Nightmares don't appear to have any therapeutic value in terms of helping us cope with the psychological conflicts and stresses of the daily lives of the dreamer. Instead they seem to contribute to our stress. Some people are continually plagued by nightmares and researchers have discovered that people with these disorders often have poorly defined boundaries—they are vulnerable and are often too open with others. Repetitive nightmare sufferers are usually very inward-oriented and have a vivid fantasy life. There are other repetitive nightmare sufferers—crime victims, soldiers, and rape survivors—that *do* know the source of their bad dreams, and their nightmares are symptoms of post-traumatic stress disorders. If their dreams don't go away with standard dream therapy, their ultimate recourse is a type of therapy called "image reversal therapy," as practiced by Barry Krakow, M.D., medical director for the Center for Sleep Medicine and Nightmare Treatment in Albuquerque, New Mexico, where the recurrent nightmares are reenacted and then rescripted, with a more positive narrative, while the dreamers are awake.

Sometimes nightmares can even be more extreme, though it seems some of the most frightening are reserved for children. In fact, some unlucky children suffer from a condition called night terror, a state not dissimilar to sleep paralysis, where the child victim cannot be awakened from their horrifying dreams. One of the symptoms of night terror is a sort of paralysis that puts them into a twilight state where the terrifying details of their nightmares remain visible to them, even with adults shouting at them or shaking them. Eventually the child falls asleep, and when they wake up in the morning, they have no memory of these episodes. But night terrors are like daydreams compared to the most extreme nightmare syndrome of all, and one that is almost invariably fatal.

In the USA in the early 1980s, a number of bizarre, unexplained deaths came to the attention of the Centers for Disease Control in Atlanta. Apparently a significant number of immigrant men of Asian background, particularly Vietnamese men from a specific tribe called the Hmong, were dying in their sleep under strange circumstances. The case studies seemed to indicate that the victims had all died under extreme duress while asleep. Furthermore, there was no pathological explanation for their deaths—the victims were healthy and had no underlying medical conditions.

Eventually a strange, unsettling pattern began to emerge from the epidemiological studies. It appeared that every one of these men had died from fear during the middle of a particularly terrifying nightmare! This discovery, coupled with their Asian racial origin, supplied the first appellation for the newly classified affliction, Oriental Nightmare Death Syndrome. Typically, 99 percent of victims were male, between the ages of twenty and forty-nine, and of Asian origin (Southeast Asian, Hmong, Hawaiian, Japanese, and Filipino). The syndrome also seemed to have a genetic element, as 18.3 percent of the victims also had brothers who had succumbed to the same syndrome. The attacks would almost always strike during the REM phase of the victim's first sleep cycle of the night.

Most of the initial research on Sudden Unexplained Nocturnal Death Syndrome, as it is now called, was conducted on Hmong immigrants in the United States in the 1980s and '90s. Many Hmong, a tribal people from North Vietnam, immigrated to the United States after the Vietnam War. Death from SUNDS began to be reported among the Hmong at a rate of ninety-seven deaths per hundred thousand Hmong males, with the first reported U.S. death occurring in 1977. The Hmong who survived these attacks reported that a nocturnal, suffocating spirit would crawl onto their chests at night and suffocate them. In the July 1983 issue of the *American Journal of Orthopsychiatry*, Dr. Joseph Jay Tobin recounted how one of his patients who had survived several attacks said that, during each attack, something came into his bedroom, sat on his chest, and tried to crush the air out of his lungs. In one nightmare it was a cat, in another a dog, and once it was a woman. Death from SUNDS is caused by ventricular fibrillation (a heart attack in the lower left ventricle striking in the absence of any cardiovascular disease) brought about by extreme terror. The medical explanation is that breathing irregularities and heart arrhythmias caused by REM sleep, as well as atonia, may play a role in causing SUNDS. Typically the victims are found dead in their beds, lying on their backs, with their faces frozen in a rictus of terror.

People who have witnessed victims dying from SUNDS say that their breathing becomes agonized and then they begin to make choking, gurgling, or gasping noises. Death usually follows shortly afterward. There is no known treatment. Just like children during an attack of night terrors, the victims are difficult to arouse, though some lucky individuals have been rescued by being splashed with water, shaken, massaged, or even bitten.

The similarity between the accounts of SUNDS survivors and various historical legends of nightmare entities is uncanny. Accounts of nocturnal beings that attempt to suffocate sleepers have been chronicled from Roman times up to the present. In the Western world they are known as incubis, or male demons that descend on sleeping persons. (There is said to be a female version of the incubus, called a succubus, who has intercourse with sleeping men.) The Chinese refer to the sensation of nocturnal suffocation as *gui ya* ("ghost pressure") while the Japanese call it *kanashibari* ("the devil stepping on the sleeper's chest"). Perhaps these demonic figures are nightmare archetypes that personify the feelings of suffocation that sometimes accompany atonia and transient episodes of sleep paralysis. But whether they are archetypes or, as some believe, manifestations of the spirit world, there is no doubt that incubi are mythic entities that transgress the border between nightmares and the supernatural. It appears that our dreams can sometimes engender monsters, beings with an existence beyond the imaginings of individual dreamers. Our fears and unconscious dreads sometimes incarnate in the real world, and sometimes, like somnambulists, there are those among us who act out these dark impulses.

9

THINGS THAT GO BUMP IN THE NIGHT—1 A.M.

From ghoulies and ghosties and long-leggety beasties, And things that go bump in the night, Good Lord, deliver us!

CORNISH PRAYER

IT IS 1 A.M., THE beginning of deep night and the first of the "wee hours," or as the Spanish say, *"tarda noche madrugada."* Day workers who went to sleep between 11 P.M. and midnight are entering their second sleep cycle while their core body temperature continues to drop. People who are still awake (unless they are night owls) are part of two groups, night-shift workers and those who can't sleep. Insomnia has many causes: too much coffee, an important event coming up the next day, but for anyone who is phobic about darkness, a spell of sleeplessness can stretch into an unbearable period of free-floating anxiety. "A sleepless night is as long as a year," according to an old Chinese proverb. And if you have nyctophobia, the fear of night and darkness, then the night can seem even longer. Every unfamiliar sound is mistaken for an intruder (or perhaps a criminally insane stalker), the family cat keeps staring at an empty chair as if somebody was sitting there. That's why Edgar Allan Poe set his sinister poem, "The Raven," at this very time, just after midnight:

> *Once upon a midnight dreary, while I pondered, weak and weary,*
> *Over many a quaint and curious volume of forgotten lore—*
> *While I nodded, nearly napping, suddenly there came a tapping,*
> *As of some one gently rapping, rapping at my chamber door.*

On the other hand, sometimes there *are* things out there, and not all of them are quiet. We've probably all been kept awake at some point by a dog barking in someone else's yard. And it's not just the noise that disturbs us; whatever shreds are left of our primal instincts tell us that the dog might be barking at something moving in the dark. (The Romans believed that dogs barked when the sorceress Hecate passed by invisibly in the night.) More often than not there's nothing there. Because dogs feel some of the same anxiety that we do at night, they bark at unfamiliar noises and scents, which for them are more plentiful in the darkness. Sometimes, like us, they just get lonely and like to know they have canine company. Dogs conduct nocturnal conversations with each other, sometimes at quite a distance. Whenever a dog seems to be barking for no reason, if you listen carefully enough, you can often hear a faint reply. Nevertheless there is something very unsettling about a dog's howl. According to an old superstition, a dog howling in front of a house at night is a bad omen for whoever lives inside. Dogs' howls evoke our primordial memories of their wolf ancestors—and most of us harbor a deep, almost instinctive fear of wolves. The following excerpt from Bram Stoker's *Dracula* plays on exactly that fear. As Renfield approaches Dracula's castle for the first time, he reports:

> *"Then a dog began to howl somewhere in a farmhouse far down the road—a long, agonised wailing, as if from fear. The sound was taken up by another dog, and then another and another, till, borne on the wind which now sighed softly through the Pass, a wild howling began."*

Cats also make hair-raising sounds at night, particularly when they mate or fight. Their demonic caterwauls sometimes sound like tormented banshees. The most amazing nocturnal cacophony of domestic animal sounds I have ever heard was on the island of Tobago. At least once every night, when the moon was full, and sometimes twice on those nights, every animal in Bucco village would start calling simultaneously. Usually a lone dog or a single rooster would start the chorus, but then, instantly, everything else, goats, chickens, cats, other dogs, and roosters, would join in. It was as if they were all a little nervous, waiting to be triggered into a sonic pandemonium by any sound or movement, and the noise they made was impossible to sleep through, though it didn't last long. As mysteriously as it began, the clamor would subside into silence, leaving only the whispering of the trade wind in the palms.

Gothic Nights

Our attitude to night underwent a revolution in the nineteenth century, as a direct result of two developments: the spread of gas lighting through the cities and Romantic literature, particularly the English Romantic poets. They embraced night, singing its praises in such poems as Longfellow's "Hymn to Night," and Lord Byron's "Hebrew Melodies," where he equates night and feminine beauty: "She walks in beauty, like the night." But the Romantic poets had a split agenda for darkness: at the same time that they began to reclaim night from the superstitions and fears that it had inspired during the centuries of the Dark Ages they also initiated a genre of literature that took the fear of darkness to new, previously unequaled levels. It makes sense, psychologically, that when society began to lose its nocturnal-siege mentality and night became a playground for Romantic souls, the fear of night would become, in a sense, a recreational fear, a titillation playing on the instinctual, but now redundant, phobia of night. In other words, as gas lighting spread through the great cities, and the curfew of darkness was pushed back, the Gothic writers and then Romantic writers were quick to use the standing pool of the recently unemployed nocturnal fears and give them a new purpose, wholly sensational. They gave us engaging, fantastic, and abstract terrors, and these funneled the fear of darkness away from the real night, which then became a more benign amusement theater where new freedoms were possible—at least for those whose innate fear of darkness wasn't that strong to begin with. The early Victorian era became the last period in which the night was regarded as intrinsically bad, as host to harmful "night vapors." With the rationalism of science, which rejected wholesale the notion of night as being unhealthy, coupled with the romantization of night by the English symbolist painters such as Edward Burne-Jones and the Romantic poets, people were freed from the tyranny of dread that had haunted humans since time immemorial. Night outdoors became a new environment to be savored, as Robert Louis Stevenson wrote: "Night is a dead monotonous period under a roof; but in the open world it passes lightly, with its stars and dews and perfumes."

The Romantic Gothic literary genre was hatched on the shore of Lake Geneva in the summer of 1816. Lord Byron had rented a splendid lakeside residence, the Villa Diodoti, for the summer, partially to escape the rather Byzantine web of mistresses and affairs he had spun in England. Lord Byron had invited his physician, Dr. John Polidori, who was a bit of a literary dilettante, to stay with him. The Villa Diodoti itself had an illustrious literary pedigree, having sheltered John Milton as well as French authors Voltaire and Rousseau. In June, Lord Byron invited his friend Percy Bysshe Shelley and Shelley's nineteen-year-old bride, Mary Wollstonecraft Shelley, to stay a few days at the villa. The Shelleys arrived on June 15, with Mary's stepsister, Claire Clairmont, accompanying them. Percy had just abandoned his wife and eloped with Mary Wollstonecraft, and they were escaping the outrage their liaison had caused in England. Claire, who was even younger than Mary, at just eighteen years, had accompanied them to try and rekindle her affair with Lord Byron. (Ironically, she was one of the English "complications" that Byron was trying to escape from in Switzerland.) That night Byron held a sumptuous reception where the wine flowed freely and, according to some accounts, the guests sampled some exotic inebriants that included laudanum and opium. The mixture of temperaments was heady and intoxicating in and of itself, but combined with the sexually charged atmosphere and the effects of visionary drugs, the evening became, according to all accounts, tumultuous. The company retired late, though according to some the night's excesses saw the light of dawn. The Shelleys intended to leave the day after, on the sixteenth, but a series of extraordinarily violent electrical storms that afternoon put their plans on hold. That evening there was a terrific storm, and after dinner they all read ghost stories to each other. Just before the night ended, Byron challenged everyone present to write an original ghost story, and to take the Gothic form, pioneered by such writers as Anne Radcliffe, to new heights.

All accepted the challenge, but none took it more seriously than Mary Shelley. A few nights later, while she was still at the Villa Diodoti, she had a dream in which she saw a man kneeling beside a human form that he had assembled out of pieces of other humans. When she awoke the next morning, she wrote the lines that begin chapter four of *Frankenstein*, "It was on a dreary night in November . . ." Mary Shelley's *Frankenstein*, published in 1818, was the first and most famous product of the literary pact of the night of June 16. She later recalled that her intention with *Frankenstein* was to write "a story that would speak to our mysterious natures," but it did much more than that: with its fusion of science and mystery, it anticipated the science-fiction genre of the twentieth century by almost eighty years. Dr. Polidori was second to publish, in 1819, with his story "The Vampyre" in the *New Monthly Magazine*. Byron and Percy Shelley also wrote stories, but they were not as successful as the others. The two themes of Mary Shelley's and Dr. Polidori's books—scientific monsters and vampires, respectively—became, along with the traditional ghost stories of the popular Gothic form, the twin pillars of modern horror.

The Gothic genre that the Romantics launched with *Frankenstein*, and later, Bram Stoker's *Dracula*, became a sort of tonic, a patrician pastime to "to curdle the blood, and quicken the heart," as Mary Shelley described it. This new literary genre crossed the Atlantic and quickly took root in America, where a young Edgar Allan Poe read the English authors avidly and began to write stories himself. Eventually he introduced readers to new, more psychological horrors, with stories like "The Fall of the House of Usher" and "The Pit and the Pendulum." A master of atmospheric horror, he once described himself, no doubt tongue in cheek, as "insane, with long intervals of horrible sanity." The lives of the Gothic authors were becoming part of their mystique, and real life did sometimes seem to imitate the Gothic genre, which seemingly set the stage for the first serial killers, like Jack the Ripper. In the Gothic age, even crime seemed to become irrational, the product of twisted psychologies and fiendish motives.

Perhaps the classic period of Gothic-Romantic literature occurred in the last two decades of the nineteenth century. It started with Robert Louis Stevenson publishing his own remarkable contribution to the literary tradition, *The Strange Case of Dr. Jekyll and Mr. Hyde*, in 1885. Stevenson elaborated on the theme of the mad scientist introduced by Mary Shelley. But what was new about his book, and what linked *The Strange Case of Dr. Jekyll and Mr. Hyde* to both the vampire genre and to the later werewolf films of the twentieth century, was its theme of nocturnal transformation, the deeply held psychological intimation that we become something "other" at night. Stevenson's book played out this phobia of "self as stranger at night" with a scientific metaphor that allowed his readers to experience their irrational fear of the unconscious self and their basic fear of the dark just at that moment in history when urban darkness was on the wane. The timing was no coincidence. By 1885 the first electric lighting systems were turning city twilight into daytime brilliance. In addition, a new spirit of psychological exploration of human behavior was emerging, popularized by the likes of the French psychologist and hypnotist Jean Martin Charcot, who was already making extraordinary claims about the hidden side of the human mind. It was in this zeitgeist that the uncontested masterpiece of vampire literature, Bram Stoker's *Dracula*, was published in 1897, only two years after Sigmund Freud had "discovered" the door to the unconscious mind in Vienna. Although Stoker did have two precedents to base his book on, Dr. Polidori's and one by a German author by the name of Johann Ludwig Tieck, his *Dracula* wholly recast vampire folklore for all subsequent generations and the ultimate mythology of the undead was revealed in all its malefic splendor.

A year after *Dracula* was published, American writer Henry James published his novella *The Turn of the Screw*. This was a cleverly structured ghost story that initiated the "evil child" theme which itself became a threnody of twentieth-century horror, as represented in such films as *Village of the Damned* (1960) and *The Omen* (1976). In fact, the twentieth century ushered in a Gothic renaissance, and less than two decades after *Dracula* and *The Turn of the Screw* were published, the first "horror" movies began to garner wide audiences. Within another two decades, film became the dominant entertainment, and watching spine-tingling horror in a dark movie theater turned into a rite of passage. Films like *The Mummy, Nightmare on Elm Street, The Exorcist*, and *Halloween* took place almost exclusively at night, a darkness amplified by the twilit ambience of the movie theater itself. (People watching horror movies at home on video perpetuate this ritual by turning off all the lights in order to invoke the fear that darkness brings.) But within the genre of horror movies, it was the vampire legend that dominated the field in myriad productions including, among many others, the classic *Dracula* in 1931, *Horror of Dracula* in 1958, and more recently, Francis Ford Coppola's 1992 film, *Bram Stoker's Dracula*. With vampires, werewolves, ghosts, and others, the twentieth century engendered a panoply of inimical beings.

Children of the Night: A Natural History

Listen to them. Children of the night, what music they make.
DRACULA (1931)

The characteristic features of the vampire are an amalgam of literary innovation and folkloric traditions and follow an archetypal logic. Vampires have many powers. First of all, because they are "undead," they are immortal. The condition of being "undead" is a special state of existence that requires fresh human-blood meals to remain active— otherwise vampires go into a sort of hibernation. With enough blood meals, they can become younger, reversing the aging process. Not only does the blood of living humans fuel their unnatural longevity, but also their bites have the ability to transform their victims into vampires themselves. If a vampire bites us, then, as Professor Van Helsing warned in Bram Stoker's *Dracula*, we become "foul things of the night" like the undead themselves.

Vampires also have unnatural strength and intelligence. They do not cast a shadow (one of the age-old marks of someone whose soul has been taken by the devil) nor do they cast a reflection in a mirror. Vampires have command over ghosts and the elements, including storms, fog, and thunder. A vampire can also control animals, particularly rats, owls, bats, foxes, wolves, and moths. In addition vampires have the ability to make themselves small or invisible or transform into a wolf or a bat. A vampire can even turn him- or herself into a mist and slip under a door. But vampires are also famously vulnerable. During the day, they must sleep in a secretly located coffin that rests within their native soil. Garlic and crucifixes repel them and they cannot enter a home unless someone in the household has invited them. Two things can kill them, a stake driven through their heart and the light of the sun.

The vampire myth holds a powerful fascination for people all over the world. Vampires appear in the folklore of cultures as disparate as Africa, Greece, Canada, and Ireland. They figure in dozens of films, hundreds of books, short stories, poems, as well as a half dozen operas, many paintings, and even in the venerable children's show *Sesame Street*. Part of the attraction is that the vampire has a privileged state of existence that transcends that of common mortals. But in exchange for immortality, the vampire pays the price of a twilight existence, forever unable to go out by day, to soak up the light of the sun. They have exchanged their health for a perpetually anemic and febrile condition that is the cost of their deathless existence. In a sense they have become emissaries of death, and of nightmares. Because they sleep in coffins, vampires incarnate our primal association between night and death, an association as old as humanity. As Renaissance poet Samuel Daniel wrote, "Care-charmer Sleep, son of the sable Night, Brother to Death, in silent darkness born." The vampire represents both death and blasphemous resurrection, a deadly symbiosis.

Vampires also personify what we both fear and desire in the night. At night the margin between life and death becomes indistinct, and in this zone cohabited by the living and the dead, the vampire is master. Vampires control that which we cannot—death and night— while enjoying that which we wish for, forbidden carnal pleasures. It appears that the myth of the vampire is not just an allegory for our ambivalence about the night, it is also an expression of the secret lusts that night releases. After all, the vampire is a highly erotic being with charismatic powers. As Freud wrote in his book *Three Contributions to the Theory of Sex* in 1910: "When the more normal aspects of sexuality are in a state of repression there is always a tendency to regress towards less developed forms. Sadism is one of the chief of these, and it is the earliest form of this—known as oral sadism—that plays such an important part in the Vampire belief." We live vicariously through the vampire, and we identify with vampires' mesmeric powers, their ability to both enslave and seduce.

The synthesis of sexuality and darkness is one of the most compelling aspects of many nocturnal myths and practices. As we've already seen, incubi and succubi have a sexual agenda, and night's deeply sexual nature is also acknowledged by witches' covens. In their nocturnal rituals, the male witch holds a dagger while a female witch holds a cup of wine into which, with post-Freudian sangfroid, the male dips the tip of his knife. This ritual is a form of pagan fertility rite, and more extreme versions of the same erotic invocation are practiced by satanic cults. It seems that the "horned one" has more than desecration and the possession of men's souls on his mind.

The notion of satanic possession seems to be the ultimate representation of the fear of night. Vampires, werewolves, and ghosts, on the other hand, are more like nocturnal nuisances, morally corrupt and evil but not directly under the devil's command. At best they are associates of the devil. The most salient, mythical features of satanic possession were well documented in William Friedkin's 1973 film *The Exorcist*. Though many of the scenes in *The Exorcist* take place during the day, the nighttime scenes are the most stomach-knotting. It is at night that the devil enters the body of the little girl played by Linda Blair, and it is at night that the final exorcism, with all its baleful effects, is carried out.

In the demonic possession myth, whose history predates Christianity by thousands of years, a demonic entity or inimical spirit enters the body of a living person and uses the body as an earthly vehicle for its own ends. In a sense it is a reverse of the vampire myth, because in demonic possession the undead spirits must find a host body—they have none of their own. Furthermore, they cannot "infect" a new body unless they migrate out of the one they are already possessing. That is why a possessing spirit is more desperate than a vampire, and much more chilling.

The idea of demonic possession acknowledges not just our ambivalence and insecurity about night, not just the changes in our own psychology that come with the advent of darkness, but it also acknowledges our deepest fear, that of our own insanity. With sleep we become the unconscious victims of our own demented hallucinations, visions that sometimes trouble us the next day, or for days afterward. We are the unwilling guinea pigs of the psychological effects of the circadian neurohormones that are released at night into our bodies. As a result, not only does our psychology change in the evening with the release of hormones like serotonin, melatonin, and cortisol, but also, when we sleep, we are the paralyzed victims of our own psychotic hallucinations. As Harvard University professor and sleep researcher Allan Hobson wrote, "Were it not for the fact that we are asleep when they occur, we would be obliged to say that our dreams are formally psychotic and that we are all, during dreaming, formally delirious and demented." Like Dr. Jekyll and Mr. Hyde, we really do become someone else at night; our transforming elixir is a cocktail of circadian hormones, and our monstrous behavior is enacted in our dreams. As Jung wrote in *The Psychology of the Unconscious* in 1953, mankind "has a shadow side to his nature which is not just made up of small weaknesses and blemishes, but possesses a positively demoniacal impetus . . . a delirious monster." It is a subconscious awareness of our own instability, our potential for madness, that underscores our nocturnal insecurities and that provides a rich ground for these metaphors of possession, vampires, and hauntings with which we people the night.

When night / Darkens the streets, then wander forth the sons / Of Belial, flown with insolence and wine.

JOHN MILTON

The werewolf myth is positioned somewhere halfway between the vampire legends and Dr. Jekyll's monster. Unlike vampires, lycanthropes can transform into only one animal, the wolf. Within this limited palette, they can either grow coarse hair and wolflike fangs, like Lon Chaney in the 1941 film *The Wolf Man*, or they can simply behave like a wolf, as Jack Nicholson did in the remarkably understated 1994 film *Wolf*. There are as many degrees of transformation as there are variations on the werewolf legend. The werewolf's transformation is contingent on one additional factor besides night—a full moon. Full moons have a long tradition of inducing madness and the werewolf legend embodies this notion: the "wolf within" is literally moonstruck. But werewolves share many qualities with vampires: they can "infect" others, turning them into werewolves, and like vampires the business of werewolves is the consumption of flesh. Also like the vampire are the werewolf's vulnerabilities: it is repulsed by wolfbane, a plant, and it can be killed by a silver bullet.

Vampires and werewolves, in fact, seem to have descended from a common ancestor and it is only in the last few centuries that they've separated into two distinct species. It has been suggested that both the vampire and the werewolf myths were spun out of rumors surrounding individuals suffering from a rare genetic disorder called porphyria. During a 1985 conference of the American Association for the Advancement of Science, David Dolphin, a biochemist, pointed out that the main symptoms of porphyria—a blood disorder that results in severe photosensitivity, the formation of thick facial and body hair, and red teeth and fingernails due to the deposition of porphyrin—were all associated with werewolves. As the disease progressed, the behavior of the afflicted individual was often accompanied by symptoms of hysteria, delirium, and bipolar psychosis. Some ethnologists believe that vampire and werewolf legends had their origins in the hearsay that grew up around early victims of porphyria, combined with the folk history of the fifteenth-century Transylvanian ruler Vladislav Tepes, more popularly known as Vlad the Impaler, who was reputed to have killed more than forty thousand people, often by impaling them. If, as legend has it, the ghosts of murdered people return to haunt the place where they met their demise, then Vladislav Tepes's castle must have hosted a towering, swirling vortex of vengeful spirits.

> *I am half inclined to think we are all ghosts, Mr. Mander. It is not only what we have inherited from our fathers that exists again in us, but all sorts of old dead ideas and all kinds of old dead beliefs and things of that kind. They are not actually alive in us; but there they are dormant, all the same, and we can never be rid of them. Whenever I take up a newspaper and read it, I fancy I see ghosts creeping up between the lines. There must be ghosts all over the world. They must be as countless as grains of the sands, it seems to me. And we are so miserably afraid of the light, all of us.*
>
> HENRIK IBSEN

Ghosts are night's children also, though they are very different from their undead siblings. Ghosts are completely dead—they are disembodied spirits that have risen to roam through the night. They do not have the single-mindedness to pursue their strategic goals in the darkness and it is very rare for ghosts to have enough power to successfully invade a living person. They are inchoate, somewhat aimless entities that exist in a parallel world not completely contingent with our own. The night they inhabit is a netherworld of phantasms and visions, of uncertainty where dreams are mistaken for apparitions and apparitions are mistaken for dreams. When Shakespeare's Horatio thinks he sees Hamlet's father's ghost atop the castle wall "In the dead vast and middle of the night," he can't really be sure whether he has seen him or not. Ghosts are natural inhabitants of our nocturnal ambiguity, moving with uncertain ease through the half-dreamed shapes of fears and memory, and they have varying abilities to terrify us. Some seem harmless, content to endlessly repeat a seemingly innocuous episode of their previous existence, while others are entirely sinister, the icy, hair-raising heart of night terror. Ghosts are almost always associated with a particular place—a house, a tower, or an old hotel—and usually return to that location in order to complete a task they were unable to finish while they were alive. Some famous haunted houses include Saint Anne's in Cheltenham, England, as well as the Tower of London. The Whaley House in San Diego, California, is one of the most well-known haunted houses in the United States, and in Japan there are many reputedly haunted natural locations, particularly ponds or sections of forests. (Tibetan Buddhist priests can see if you have "hungry ghosts" surrounding you like invisible parasites. They believe that covetous and greedy people are reborn as hungry ghosts who, although they are tormented by great thirst and hunger, have tiny mouths and knotted throats and so cannot eat or drink. A person who is materialistic in this life sometimes attracts hungry ghosts, who are drawn to them like kindred spirits.) Ghosts are often the tormented spirits of victims of great violence or the spirits of those whose lives were visited by tremendous sadness. In addition, ghosts seem to represent life tasks we haven't finished, as well as the will to live beyond death. Because ghosts are thought to prove that the soul survives the death of the body, that there is indeed an afterlife, albeit a strange half-life, ghosts are emissaries of a secular universe. They contradict monotheistic religions and are seemingly untouched by heaven or hell.

Night is when we remember those who are gone and those whom we have loved and lost, though they may still be alive. As James Joyce wrote in *Ulysses*: "What is a ghost? Stephen said with tingling energy. One who has faded into impalpability through death, through absence, through change of manners." Ghosts are those whose afterlife is so insubstantial, so tenuous, they can only exist in the nurturing mezzotint of darkness. It's as if the power of our memory, or our longing, can conjure the image of our lost ones out of the night, but like Orpheus yearning for Euridyce, we cannot grasp them. If, in the pantheon of our own nocturnal psychology, vampires are death and sexuality, werewolves incarnate our primitive drives, while demons and monsters embody the fear of our own nocturnal transformation, then ghosts are memory and death and absence.

Some ghosts seem consigned to repeat menial tasks and are often seen at the same time of night performing the same actions. Perhaps these ghosts represent the sum of all our unconscious acts, those things we do when we are preoccupied or not fully attentive to what's in front of us. Perhaps by not living fully in the moment, we lose a bit of ourselves, building up a ghostly repository of unawareness. As if a phantom self was invoked by the sum of our various inattentions and this sad, abandoned self is the ghost we see in the night.

> *Faeries, black, grey, green, and white*
> *You moonshine revellers, and shades of night.*
> WILLIAM SHAKESPEARE

If ghosts are the most tenuous children of the night, fairies are the most beguiling. The border between ghosts and fairies is straddled by the banshee, a female fairy that is said to appear just before a death. She sits atop a fence post or a gate outside of the home of the imminently deceased and wails horrifically. The Irish practice of keening, the mourning cry that is traditionally sounded at funerals, is thought to be an imitation of the banshee's terrifying wail. If a great man or woman is about to die, more than one banshee will appear, sometimes accompanied by other dire omens, such as the arrival of a black coach carrying a coffin drawn by headless horses.

Fortunately most fairies are not as portentous as the banshee; in fact, according to legend most are shy, elfin creatures that shun humans. That's not to say there isn't something alien, almost inimical, or even pathological about the folklore regarding fairies. An infestation of fairies can spell no end of trouble for cottage dwellers, for example, though tormented homeowners do have recourse to two tried-and-true remedies for repelling fairies. Fairies can be repulsed by holy water or by turning your clothes inside out. It is only the rare fairy that is actually helpful. Among these are brownies, who will do minor chores for you at night if they are so inclined.

In Gaelic lore certain hills are hollowed out by fairies for their homes, and on quiet nights if you lie with your ear to the earth on these hills, you can hear the music of the fairies. Legend has it that you must never fall asleep while listening to fairy music, for you might not wake up. Malicious Manx fairies called Hogmen move from one hill abode to another on the night of November 11, and according to Manx wisdom, it is best to stay inside on this evening. Another reputed reason to avoid fairy hills at night is because witches often visit them.

The Irish have a rich history of fairy lore. According to some Irish legends the king of the fairies is the ancient king Finvarra, once a giant, who has shrunk over the years to human size. He and his splendid court cannot go out by light of day, but at night Finvarra often abducts beautiful women to take back to his bed. The theme of abduction runs throughout fairy legend. Sometimes fairies kidnap human babies and leave one of their own in their stead. These babies are often sickly and pale and are called "changelings."

A ring of toadstools in the grass is called a "fairy ring" and is said to mark where fairies dance at night. If by chance someone comes upon fairies dancing at night, he must never join the fairy dance. If he does, he will become a captive of fairy magic and dance for seven years, although the poor bewitched human will think he's only been dancing for one night. Mushrooms are good props for these nocturnal revels as they are the only plants that do not need light to grow. Toadstools and fungi flourish at night and literally spring up in the darkness. The queen of all the toadstools is the fly agaric (*Amanita muscaria*). It is usually red, orange, or deep yellow with a round cap sprinkled with white warts. According to legend fly agaric was taken by Vikings to sharpen their battle skills, though sometimes the hallucinogenic effects of the mushroom would trigger their infamous "Berserker" frenzies on the battlefield. (Curiously enough, this prebattle ingestion of a psychopharmaceutical foreshadowed the use of Benzedrine by the German Panzer divisions that overran Poland in 1939, allowing the Blitzkrieg to continue advancing all night long without being interrupted while tank crews slept. Amphetamines are also routinely dispensed to military night patrols and reportedly to American fighter pilots on night missions.) The use of *Amanita muscaria* by Siberian shamans has been well documented, and I have read firsthand accounts of ethnobotanists who have ingested these mushrooms. They all claim they experienced a feeling of great self-confidence and superhuman strength. The Siberian shamans reported that the mushrooms give them the ability to see great distances and into other dimensions. Perhaps the fairies' own magical powers were in some way associated with the effects of these fungi.

> *And we faeries, that do run*
> *By the triple Hecate's team*
> *From the presence of the sun,*
> *Following darkness like a dream,*
> *Now are frolic.*
>
> WILLIAM SHAKESPEARE,
> *A MIDSUMMER NIGHT'S DREAM*

Of the fairy realm only a handful are truly nocturnal, including goblins, who often inhabit mines, according to English and German legend. In a poem called "Goblin Market," about two sisters, Christina Rossetti relates how one of them, Lizzie, is ravished by goblins at night in the forest. Afterward Lizzie returns home to her sister, Laura, and says:

Did you miss me?
Come and kiss me.
Never mind my bruises,
Hug me, kiss me, suck my juices
Squeezed from goblin fruits for you,
Goblin pulp and goblin dew.
Eat me, drink me, love me;
Laura, make much of me:
For your sake I have braved the glen
And had to do with goblin merchant men.

It is plain that the goblins had effects in addition to simple bewitchment. Lizzie was so erotically charged when she came back to Laura that Laura also succumbed to the nymphomaniacal spell and began to kiss Lizzie passionately: "She kissed and kissed her with a hungry mouth." The hijacking of human passions by magic proxy is a theme throughout fairy lore and Rossetti's poem captures its essence.

Dwarves are also purely nocturnal fairies, and contrary to the dwarves in the *The Lord of the Rings* trilogy, they cannot go out by day, otherwise they turn to stone. Dwarves live inside mountains where they fashion extraordinary metallic objects. Other night fairies include the asrai, delicate female water fairies that are shade loving, if not truly nocturnal, for they melt into a pool of water if they are exposed to sunlight. Pixies are also nocturnal, mischievous fairies.

The plethora of nocturnal entities, fairies, vampires, and werewolves proves that night is a fertile medium for imagination. In the darkness we become unsure of our surroundings and so we people the night with materializations of our fears, with goblins and ghosts. Given our ancestral experiences of nocturnal creatures, some of which preyed upon us only in the night, it is probably an intrinsic part of our human identity to expect a wholly different class of creatures to be abroad by dark, not all of whom wish us well. This is perhaps why the stories of people transforming into animals are so compelling. Of course, night attacks by other humans have also occurred thoughout our history, and historically, with our protean propensity for remaking ourselves with masks and paint and fur, these "wild men" have added fuel to the legends of malevolent beings in the night.

But our collective mythology continually refigures itself and new legends are constantly being created, particularly in response to new technologies. During the last century, with the advent of electronic telecommunications, flight, and the nuclear bomb, a completely new set of legends was spawned. As always, night provided the perfect tabula rasa within which these myths could manifest themselves.

UFOs: Visitors from the Stars

As the archetypes of our dreams have kept pace with technology, so has our folklore. Now we have nightmares about nuclear wars and plane crashes, and in the last half of the twentieth century, a new mythology of alien technology has arisen. The onset of this collective myth was sudden. Starting in 1947, reports of strange lights flying through the night skies began to flood into police stations, television news studios, and Air Force bases. Something new, something sinister and alien, was out there, and we were being watched, even abducted.

The golden age of UFO sightings took place during the 1950s, predominantly in the United States, although there were other sightings reported from around the world. By far the largest category of sightings was of luminous objects at night. Usually these "crafts" were saucer-shaped, though many were cigar-shaped or triangular. Sometimes radar operators claimed to be able to track these objects and there are confirmed reports that at least on one occasion jet fighters were scrambled from an American Air Force base to intercept them. UFOs were said to make extraordinary flight maneuvers, zigzagging, hovering, and accelerating out of sight at terrific speeds. A smaller number of observers claim they saw UFOs land, and of these an even smaller percentage reported seeing "humanoids" in the vicinity. Invariably these humanoids were described as "little men" and their reported features have been so widely disseminated in the media that almost everyone is familiar with their triangular-shaped heads and large, almond-shaped eyes.

Of all the myths and legends that have come out of UFO folklore, the most psychologically compelling stories have been centered on abductions, mutilations, and crop circles. Although depictions of abductions in films almost always have a rapturous, visionary look to them, with the victims rising into a beam of intense, heavenly light in the middle of the night, their bodies seemingly transfixed in a position of religious ecstasy, people who claim to have been abducted in real life report harrowing experiences on board UFOs. They report being subjected to bizarre experiments or sexual assaults and surgical mutilations. These mutilations involve alien medical procedures that often focus on the human reproductive system. This theme was echoed in the rash of farm animal mutilation reports that peaked during the 1980s in the American West—where cows, sheep, and other farm animals were claimed to be found dead or injured with what appeared to be surgical incisions. Upon examination it was discovered their genital organs or even particular internal organs had been neatly and precisely removed. The alien surgical theme waned during the early nineties when it was supplanted by the more benign phenomenon of crop circles. (It seemed that the aliens had moved from sadism to more aesthetic activities.) Like toadstools or fairies, most of these rural phenomena seem to take place at night, and the results of the nocturnal activities of UFOs are only discovered the morning after.

Even I have spotted enigmatic lights in the night sky and wondered if they were representative of a more sophisticated civilization. But why do we need to believe in extraterrestrials? Perhaps we intuit, on an unconscious level, that we are not fully evolved, and that, at some point in the future, our species will perhaps look back at this stage of our evolution with pity and horror. Certainly UFOs represent our fears, but they also symbolize our hope for transcendence. Perhaps UFOs are an answer to our existential loneliness as a species. On some deep, collective level we are desperately alone, and when we realized the vastness of the universe four hundred years ago, our concomitant inconsequentiality made us even lonelier. We want company. Maybe that's why we're trying to communicate with anything that could be intelligent. Unfortunately our attempts to converse with animals seem doomed; we would love it if dolphins or elephants or even chimpanzees could be taught to speak, and to some extent they can, but not well enough to relieve our intense species loneliness. We might even eventually build an artificial intelligence to keep us company, but until then, many of us will look to the stars at night and hope for contact with someone else who can make the vastness of the universe a little less lonely.

In *Close Encounters of the Third Kind* (1977), Steven Spielberg perfectly captured the romance and excitement of visitors from the stars. The movie is filled with magical, starry nights populated with fairylike UFOs. In one sequence a police car gives chase to a mischievous UFO, lending a sort of ironic credence to the famous astronomer Fred Hoyle's observation, "Space isn't remote at all. Its only an hour's drive away if your car could go straight upwards." But the climax of *Close Encounters of the Third Kind* surpasses even the nocturnal grandeur of *E.T. The Extra-Terrestrial*. The ultimate visitation occurs at Devils Tower, in northeastern Wyoming, on a warm, summer night. The evening display begins with the flying saucers disguising themselves as a constellation, but a constellation that suddenly begins to move in the night sky. We are about to get a visit. Then the "mother ship" begins its descent to earth in a beautiful cinematic sequence climaxing a movie full of marvelous night skies. Spielberg reclaims and refigures the mythology of UFOs, wresting it from fear and instead projecting a benevolent night realm where the stars are full of mystery, magic, and ultimately, our evolutionary destiny.

The Heart of Darkness

"Who knows what evil lurks in the hearts of men?" as The Shadow used to intone at the beginning of the eponymously named forties radio show. Now our fear of the dark is more vestigial than pragmatic, though it remains somewhat active. Even if very few of us believe in the old legends of vampires and fairies, or even the new myths of alien abductions and animal mutilation, we'd be wrong to assume there is nothing to fear from the night. On the contrary, as Charles Baudelaire wrote in *The Politics of Darkness*: "Here is the charming evening, the criminal's friend; It comes like an accomplice, with stealthy tread." Sometimes there really are ghouls that stalk the night. Night has provided cover for murderous cowards and psychotic rapists. The infamous Argentine death squads stalked their *disparados* after dark. In fact, there is no predatory animal more dangerous than men with weapons, and there is no more dangerous time for them to be on the loose than at night. The primary reason we instinctively fear night is because that is when evildoers commit evil deeds. What remnants we have of our forebears' primordial fear of animal predators are nothing beside the very real, very recent fear of our own kind. The stars have twinkled over depravity and mass slaughter, and the moon has illuminated many nocturnal battles.

A recent, chilling example of murderous ghouls stalking the night was the infamous Charles Manson and his sociopathic harem called "the family." They would drive into Los Angeles from the desert at night, get high on LSD, and sneak into large homes in Hollywood while the owners were asleep. Manson called these sinister visitations the "creepy crawlies," and Sharon Tate's grisly murder was the culmination of a series of home invasions. LSD also had the effect of dilating the Manson followers' pupils, giving them excellent night vision. For real thrills they would creep into bedrooms and stand over the beds of the sleeping owners, bringing their faces within inches of those of their unconscious victims. Although there was another murder linked to their midnight ramblings, it is a miracle that the Tate slaughter hadn't been preceded by even more murders—it was only because some sleeping residents never woke up that their lives were spared.

> *You may do in the dark*
> *What the day doth forbid.*
> *Fear not the dogs that bark;*
> *Night will have all hid.*
>   THOMAS CAMPION

Night has been the traditional shelter for revolutions, military attacks, freedom fighters, and various insurgencies but it has also covered pogroms, suppressions, and vigilantes. The lynch mobs of the American Wild West rode at night, as did the Ku Klux Klan, the white supremacists who launched their evening terror raids in the south and whose ominous burning crosses were only kindled in total darkness. Espionage was historically conducted under the cover of darkness, and great treachery has often been launched at night, in legend and in history. The warriors hidden in the belly of the Trojan horse emerged at night, and all Americans remember Paul Revere's historical midnight ride. Regretfully, some of the most heinous acts ever committed have been at night, including the infamous Kristallnacht of November 10, 1938, where so many

German-Jewish shops and businesses were vandalized that the streets were filled with glass. But even more terrifying were the horrific nights of Nazi atrocities committed against Jewish families in the Warsaw ghetto.

The Second World War also marked the beginning of night bombing and nocturnal air battles. For citizens in large European cities during the Second World War, nothing was more chilling than the sound of air-raid sirens in the middle of the night. During the London Blitz, death and fire rained down out of the darkness from droning fleets of German bombers. Then came the retaliatory night raids by the Allies over Germany, including the infamous firebombing of Dresden. During the latter stages of the Second World War the German Luftwaffe developed a fleet of highly specialized night-fighters consisting of modified twin-engine Junkers with radar antennae bristling from the nose of the plane like the geometric whiskers of a mechanical nighthawk. The JU 88s also carried upward-firing cannons called "Schrage Musik" permitting them to fly underneath their targets without being seen and fire into the bellies of Allied bombers.

More recently in the two Gulf Wars, American stealth fighters and bombers have flown many night missions. The F-117 Nighthawk, America's premier stealth fighter-bomber, is not designed for nocturnal dogfights like the JU 88s of the Second World War Luftwaffe. It is an attack craft that, due to its radar-reflecting, angular shape, appears to have a profile no bigger than a seagull on enemy radar. The F-117 Nighthawk is a sinister, flying black triangle, and like a true denizen of the night, it can see in the dark with infrared targeting systems and deliver its missiles and bombs while invisible to both eye and radar. In fact, over the last few decades, the entire U.S. military has become primarily a night-fighting force, using the same advantages that nocturnal predators use—night vision, surprise, and strength—to subdue their enemies.

But before the advent of night vision and stealth technology, many battles were fought of necessity in the dark of night. During the American Civil War, Walt Whitman witnessed a nocturnal battle that he later described in his journal: "The night was very pleasant, at times the moon shining out full and clear, all Nature so calm in itself, the early summer grass so rich, and foliage of the trees—yet there the battle raging, and many good fellows lying helpless, with new accessions to them, and every minute amid the rattle of muskets and crash of cannon . . . the quick flaring flames and smoke, and the immense roar—the musketry so general, the light nearly bright enough for each side to see the other." Later he visited the scene of the battle and afterward wrote a short poem "Look down, Fair Moon":

> Look down, fair moon, and bathe this scene;
> Pour softly down night's nimbus floods, on faces ghastly,
>   swollen, purple;
> On the dead, on their backs, with their arms toss'd wide,
> Pour down you unstinted nimbus, sacred moon.

Aside from the obvious contrast between the grisly reality of the slaughtered soldiers and the peaceful, soft light of the moon, Whitman's poem gets at the deeper conjunction of nature and mankind. In a sense, the horrors that we humans have visited upon each other by night are part of the overarching reality of nature, and perhaps historically unavoidable. Yet, aside from these real and fortunately rare dangers, we are the authors of our own night and most often it is only our fear that populates its twilit recesses with demons and ghouls.

10

STARGAZING: THE MYTHOLOGY AND SCIENCE OF THE NIGHT SKY—2 A.M.

The immense deserted night set up its formation of colossal figures that seeded light far and wide. Aldebaran trembled, throbbing far above, Cassiopeia hung her dress on heaven's doors, while the noiseless chariot of the Southern Cross rolled over the night sperm of the Milky Way.

PABLO NERUDA

IT IS 2 A.M. AND most people are dreaming. If dreams leaked out of sleepers' heads then, the night would be saturated with them. Perhaps that's why so many writers work late at night, with all those dreams around to pluck out of the air. The core body temperature of the sleepers continues to drop while their cortisol levels rise. These circadian cycles that pulse in everyone's body are hard to reprogram— even hardened night-shift workers will feel a chill in the small hours, not to mention an insistent, if not narcoleptic, urge to nap.

On the northeast shore of Lake Ontario, the city of Toronto is very quiet now, and even downtown many streets are deserted. The only activity other than in emergency wards and police stations is in the entertainment district, and even that is beginning to close down. Last call has sounded in most of the clubs and bars. Soon their patrons will be hailing taxis or fishing for their keys in parking lots. Most will drive straight home, though a few will stop at all-night convenience stores, or the handful of fast-food outlets that cater to the late-night crowd. A final pulse of cars trickles back into the residential sections of the city and the bedroom communities just beyond the suburbs.

It is a clear night and the stars are bright, especially in the country. To the north the Big Dipper glitters high in the sky while Andromeda is beginning to disappear below the western horizon. Many stars are familiar, such as the three bright stars that form the belt of Orion, but the most famous star of all, the star of stars, is the North Star, also known as Polaris. It is in the constellation Ursa Minor (the Little Dipper) and is located at the end of a line inscribed by two pointer stars that make up the outward edge of the dipper in the constellation of the Big Dipper, otherwise known as Ursa Major. Betelgeuse is perhaps the second runner-up, being a fairly well-known star in the Orion constellation, especially after it was featured in the film *Beetlejuice*. The Evening Star is also renowned, especially in literature, although it's a planet (Venus) and not really a star at all. Proxima Centauri is familiar to amateur astronomers as the nearest star—it is only 4.22 light-years, or forty trillion kilometers, from earth (though it would take sixty thousand years for our fastest space vehicles to get to it)—but in terms of overall fame, it doesn't touch Polaris.

The primary reason that Polaris is the best known of stars is that all the other stars appear to circle it. It is at the center of the projected axis of the earth into space, which is why it is also called the Pole Star. But Polaris wasn't always the Pole Star; throughout the millennia other stars have taken its place. This wandering designation of the Pole Star is due to a special irregularity in the revolving motion of our planet called *precession*, which is the motion we see in a spinning top that is slowing down and beginning to wobble. Because the earth turns on its axis like a gyroscope or a spinning top, it also precesses like one. That is to say, the earth's rotational axis moves around in a small circle as it spins, only it takes twenty-four thousand years to inscribe one of these circles. Five thousand years ago, during the first classic age in Egypt when the Great Pyramid of Giza was being built, the Pole Star was Thuban, in the constellation Draco. (Five hundred years after Giza was built, when the Pole Star was still Thuban, the

Great Pyramid of Cheops was constructed with a secret, angled passageway that descended deep into the north side of the pyramid. The angle of this passage was perfectly aligned with Thuban, which could always be seen from the very bottom of the shaft at night when it was open.) Two thousand years ago, when Pompeii and Herculeneum were buried by Vesuvius, the Pole Star was Kochab in Ursa Minor (the Little Dipper). In the future another star will dethrone Polaris, but that coup is still hundreds of years away.

The Night Sky

The glittering pageant of the stars conveys a profound sense of our insignificance in a universe that is incomprehensibly huge. But it's not just the size of space that awes us; it is also the fantastic objects that populate it: quasars that outshine whole galaxies; neutron stars whose atoms are packed so densely a tablespoon of one of them would weigh thousands of tons; pulsars, the gyroscopic cinders of supernovas spinning so fast that their x-ray emissions click like a metronome over the speakers of the radio telescopes that eavesdrop on them; colliding galaxies; and giant black holes swallowing streams of stars. There are also colossal red giant stars, blue dwarfs—and somewhere, perhaps, alien civilizations of unknowable complexity—and all of this so extraordinarily far away that, even at the speed of light, it would take millions of years to get there, and "there" would be transformed beyond all recognition by the time we arrived.

The night sky is a window into an incredible, dizzying abyss that drops away in every direction we look. There is something humbling, certainly, about this spectacle, but deeper than that is the knowledge that though we are insignificant and tiny, we are nevertheless part of a cosmic reality that is miraculous. The stars at night both overwhelm and exalt us. They are destiny and judge. They care nothing about our lives and yet they shaped everything we are today. We are literally stardust, as Joni Mitchell sang in her song "Woodstock," which she wrote after she had read about a recent scientific discovery. She learned that most of the atoms that make up our bodies were new, relatively recent forms of matter that had been created in the violent, alchemical forges of ancient exploding stars called supernovas. They died so that we could be made.

The Cone of Time in the Night Sky

Starlight is pure history. The light from the star Rigel, one of the brightest points of light in the night sky, took nine hundred years to get here because it is nine hundred light-years distant. It appears to us as it was when Spain was a province of the Islamic empire during the Middle Ages. We see the galaxy Andromeda as it was when *Homo erectus* first entered Europe more than two million years ago. Distant objects called quasars are fifteen billion light-years away—their glow dates from the beginning of the universe itself.

The night sky is a mosaic of fossil light—light from the Roman era, light from the age of dinosaurs, light from before our solar system existed. Perhaps light from suns that shone on fantastic civilizations that have been extinct for millions of years. Some of the stars that seem to shine so brightly today may even no longer exist, having exploded or been snuffed out or having collapsed into black holes whose light was extinguished thousands of years ago.

When you tune your mind to the vast scale of the universe, it seems that the speed of light slows to a crawl, and the cold vacuum of space becomes a crystal jelly in which planets and stars and galaxy clusters are suspended like insects in amber. But even more strangely, the farther away a star, like the quasars that are billions of years away, the farther back in time and the closer to the big bang at the beginning of the universe that star is. Wherever we look into the night sky, we look back in time, back toward the beginning of the universe. If time inscribes a line, and the expansion of the universe is mapped out on that line, then as the universe expands, it turns the time line into a cone. When we look at the stars at night, we look down through our wide end of the cone of time toward the smaller end, the apex of creation. It is one of the great paradoxes of space, that wherever we look, the universe is smaller than it is now—yet at the same time we are surrounded by our beginning. The universe is a mystic riddle of space-time. As Hart Crane wrote, "Stars scribble on our eyes the frosty sagas, / The gleaming cantos of unvanquished space."

Nyx

And she {Nyx} bare Hypnos and the tribe of Oneiroi.

HESIOD

The stars have dazzled humans since prehistoric times—they have always conveyed a sense of magnificence and celestial mystery. The mythology of night is an embodiment of our associations with the evening sky. The Greek myths contain some of the most beautiful and curiously apt personifications of night that exist, though their stellar mythology hadn't begun when Homer first wrote about the constellations in the seventh century B.C. At that time the forms of the constellations had been mapped, but no legends or stories had been associated with them. In his *Iliad*, for instance, Homer wrote about the constellation of the Bear (Ursa Minor) without reference to any of the legends that came to be associated with it. "The bear, which men also call by the name of Wain: she wheels round in the same place and watches for Orion, and is the only one not to bathe in Ocean." Over the next two hundred years the constellations garnered their legends and stories, and by the fifth century B.C. several of the Greek legends incorporated universal archetypes. In fact, many of the themes that appear in Greek mythology recur in stories and folklore around the world. And these myths extended far beyond the representations in the stars called constellations. The intuitive pairing of death and sleep, for example, is nearly universal, and it is reflected in the literature and mythology of many nations. For the Greeks, death was personified as Thanatos, and his younger twin brother, Hypnos, personified sleep. But we are getting ahead of ourselves. The Greco-Roman family tree of night is a remarkable one, and it starts with the birth of Nyx (Night) herself.

Nyx was among the first and therefore one of the oldest and most important of Greek gods. Because of her ancient lineage, even Zeus was in awe of her and avoided displeasing her. She was the daughter of Chaos, who gave birth to the entire universe and all the original gods. These first gods were called the Protogenoi—among whom were Gaia (Earth), Tartarus (Underworld), and Erebus (Darkness) as well as Nyx. According to legend, Nyx rode a chariot pulled by two coal-black horses and drew the shade of night across the sky behind her. Nyx was able to reproduce both sexually and asexually and she gave birth to many children. Her progeny were gods also and the ones that she conceived asexually were hatched from divine eggs, like birds. These include Hypnos (Sleep) and his brother, Thanatos (Death), the Oneiroi (the Tribe of Dreams), Eris (Discord, or Strife), and Moros (Doom). And after these she gave birth to even more malefic gods and goddesses, according to Hesiod's *Theogony*: "And again the goddess murky Night, though she lay with none, bore Blame and painful Woe, and the Hesperides who guard the rich golden apples and the trees bearing fruit beyond glorious Ocean. Also she bore the Destinies and ruthless avenging Fates . . . Also deadly Night bore Nemesis Indignation to afflict mortal men, and after her, Deceit Apate and Friendship and hateful Age and hard-hearted Strife."

In Hesiod's account of Nyx's family, the children of night personify the negative nocturnal emotions that we associate with darkness. It is at night that remorse, bitterness, and depression seem most active, it is at night that we second-guess mistakes we have made during the day, when we indulge in fantasies of what could have been if we'd had the presence of mind at the time. It is at night, also, that visions of revenge seem to run rampant—and only the hopelessly rich have not spent a sleepless evening worrying over their financial woes. It is at night that the fear of death gnaws at us, that sadness over the loss of a loved one overwhelms us. This is the dark side of night and these gods are its dreaded progeny: Woe, Blame, Nemesis (Retribution), Deceit, and Eris (Strife).

But there were also some family success stories within Nyx's immaculate brood, the Hesperides for one, the nymphs that guard the beautiful gardens on Mount Atlas at the western end of the Mediterranean. The Hesperides also represent the colors of sunset. The home of Night herself is near to the Hesperides, which Hesiod describes in Theogony: "There also stands the gloomy house of Night; ghastly clouds shroud it in darkness. Before it Atlas stands erect, and on his head and unwearying arms firmly supports the broad sky, where Night and Day cross a bronze threshold and then come close and greet each other." Another successful sibling of Nyx was Friendship, the flame that burns true through the most inclement night, as well as the three Fates, though ultimately they are neither good nor bad, as the action of one undoes the other's. But of all her self-produced children none was more famous than Eros, though there is some disagreement about whether or not Eros was truly the son of Night. According to Hesiod, Eros was created by Chaos at the same time as night. It is Aristophanes who claimed that Nyx gave birth to Eros, the god of love and passion.

The only two children that Nyx had by sexual union were with the god Erebus. Compared to her immaculate births, these two seem much more benign; Light (Aether) and Day (Hemera). Hesiod refers to their carnal origin in *Theogony*: "From Chaos came forth Erebus and black Night Nyx; of Night were born Aether being the bright upper atmosphere and Day Hemera, whom she conceived and bore from union with Erebus her brother."

Hypnos

Hypnos, the god of sleep, later called Somnus by the Romans, was portrayed both as a young man with wings sprouting from his temples and as an older bearded man with wings attached to his shoulders. (It is thought that the wings attached to his temples symbolized how the mind takes flight when we dream.) Hypnos lived with his twin brother, Thanatos (Death), in the Underworld, where they played a role in many of the Greek legends. In one myth, the moon goddess Selene fell in love with a mortal, a handsome shepherd named Endymion, with whom she had fifty children. When she realized that he was going to die, she cast a spell on him, causing him to fall asleep forever so that he would never leave her. It was Hypnos who gave Endymion the ability to sleep with his eyes open. That way he could always see his beloved as she rose in the evening. This myth led to the association of sunset with Endymion, who was seen as the setting sun with which the moon was infatuated. Traces of this unrequited love between the setting sun and rising moon can be seen in a passage from Willa Cather's *My Ántonia*: "As we walked homeward across the fields, the sun dropped and lay like a great golden globe in the low west. While it hung there, the moon rose in the east, as big as a cart-wheel, pale silver and streaked with rose colour, thin as a bubble or a ghost-moon. For five, perhaps ten minutes, the two luminaries confronted each other across the level land, resting on opposite edges of the world. In that singular light every little tree and shock of wheat, every sunflower stalk and clump of snow-on-the-mountain, drew itself up high and pointed; the very clods and furrows in the fields seemed to stand up sharply."

Hypnos also played a major role in the Greek myth of Ceyx and Halcyone, when Halcyone sent her messenger, Iris, to the home of Hypnos. She asked Iris to implore Hypnos to send her a dream that would help her find her lost husband, Ceyx, who unbeknownst to her was already dead. Thomas Bulfinch's *Age of Fable* contains a wonderful description of the King of Sleep's palace when Iris came upon it:

A mountain cave is the abode of the dull god Somnus . . . Clouds and shadows are exhaled from the ground, and the light glimmers faintly . . . No wild beast, nor cattle, nor branch moved with the wind, nor sound of human conversation, breaks the stillness. Silence reigns there; but from the bottom of the rock the River Lethe flows, and by its murmur invites to sleep. Poppies grow abundantly before the door of the cave, and other herbs, from whose juices Night (Nyx) collects slumbers, which she scatters over the darkened earth. There is no gate to the mansion, to creak on its hinges, nor any watchman; but in the midst a couch of black ebony, adorned with black plumes and black curtains. There the god reclines, his limbs relaxed with sleep.

This lyrical account of the palace of Hypnos with its drowsily ominous black plumes and curtains also contains a hint of what the panorama of all possible dreams must look like, with Hypnos sleeping in their midst: "Around him lie dreams, resembling all various forms, as many as the harvest bears stalks, or the forest leaves, or the seashore sand grains."

Roman poet Ovid created the association of the god Hypnos with dreams. Previously the Greeks had attributed dreams only to the Oneiroi—the bringers of dreams. The Oneiroi were a thousand-strong tribe of black-winged daimons that swarmed out of a cave near Hades at the advent of nightfall, like bats flocking out of a cavern.

The Oneiroi shaped the dreams of sleeping people. The most skillful and adept were Morpheus, Phantasos, and Ikelos. These three created the dreams of kings and rulers, while the other Oneiroi formed the dreams of common people. Ovid embellished the original Greek myth by making Hypnos the father of Morpheus, Ikelos, and Phantasos in his poetry.

But other than shuffling paternity, Ovid followed Greek protocol by attributing the same special skills to Morpheus, Ikelos, and Phantasos that the Greeks did; Morpheus could take on the appearance of any person, Ikelos could become any animal, and Phantasos could become any inanimate object. The myth of Hypnos and the Oneiroi shows that the Greeks had a keen insight into the psychology of dream mechanisms. It reveals their knowledge, for instance, that dreams are highly mutative, and that this transformational quality disguises the real subject of a dream. In this way they anticipate Freud's notion of "dream work" twenty-five hundred years before Freud was born.

Selene/Artemis/Diana

It would seem logical that one of the children of Nyx would be the moon, but the moon had a very different maternity. Selene, the moon goddess, was the twin sister of Apollo, born to Zeus and Leto on the island of Delos. This mythic motif of twin siblings personifying the sun and moon was widespread. In Scandinavian mythology, for example, the god Balder was the day and his brother, Ho'dur, was the god of darkness. The Greeks also called Selene *Artemis*, and she was also their goddess of the woodlands and the hunt. A virgin goddess, Selene surrounded herself with virgin warriors called the Amazons. She was famous for her fierce retribution against any man who would attempt to violate her or any of her huntresses.

When the Greeks colonized southern Italy around the sixth century B.C. the indigenous Italian goddess Diana became identified with Artemis. She took on the lunar attributes as well, and by the time of the Roman republic, she presided over the night with another goddess known as Hecate, previously familiar to the Greeks as a goddess of magic. Despite the fact that Nyx was originally the ultimate goddess of night, at least to the Greeks, the Romans elevated Diana and Hecate into the dominant deities of darkness. By splitting the personification of night into two divinities, they embodied humanity's ambivalent relationship to night—Diana (Artemis) represented the glory of night, the moonlit splendor and romance of evening, while Hecate represented night's maleficent and fearful aspects. Hecate was also the Roman goddess of witchcraft and sorcery. She would wander the earth at night, invisible except to dogs, which barked when she approached. She became a mystery cult figure for the Romans, and represented the underworld and all that was connected to it.

Hecate was traditionally depicted with three heads, those of a horse, a lion, and a dog, which represented her ability to hear, see, and know all, respectively. Hecate was also believed to be the mother of a group of female vampire goddesses known as the Empusae, as well as the ruler of crossroads, where she could look in all directions. With no constellation named after her, Hecate was more of an embodiment of the earthly night. Above her the stars enacted the ancient dramas of the celestial pantheon, many of whom had earthly origins like Hecate, but who were now consigned to the heavens.

Mythologies of the Constellations

As the night advanc'd it changed its spirit and garments to ampler stateliness. I was almost conscious of a definite presence, Nature silently near. The great

constellation of the Water-Serpent stretch'd its coils over more than half the
heavens. The Swan with outspread wings was flying down the Milky Way.
The northern Crown, the Eagle, Lyra, all up there in their places.

WALT WHITMAN

Although many of the legends that personify the constellations have their origins in stories and mythology that arose independently of the forms in the stars, many others are purely interpretive and owe their narratives directly to formations of stars. An example of the first type is the legend of Andromeda, where Perseus, a purely mythic character, makes an appearance along with Medusa's head. Here the constellations are merely repositories for the names of legendary figures. A good example of the second, more intuitive mythology of the constellations is the almost worldwide interpretation of Ursa Major as being shaped like a bear, with its various attendant stories. Wherever a narrative has been suggested by the actual formations of stars or their movements relative to each other, ethnoanthropologists believe these stories are probably older and more primitive than the more recent, mythic legends that have been woven around the constellations. But whatever their origins and histories, these stories connect us with a living past that is resurrected every evening by the constellations as they parade across the night sky. Of all the constellations perhaps among the most famous are Orion (the hunter); Ursa Major and Ursa Minor, the big and little bears, respectively; Andromeda; and the Southern Cross. On November nights in the Northern Hemisphere the constellation of Andromeda stretches in a long, curved triangle almost directly over-head. Not only is it one of the largest constellations, but it was also one of the earliest to be named, with references dating back four thousand years to ancient Mesopotamia. According to Greek legend, Andromeda was the daughter of an African king, Cepheus, and his exquisitely lovely queen Cassiopeia. Cassiopeia incited the wrath of the gods by claiming she was more beautiful than the sea nymphs, the Nereids. Her punishment from the gods came in the form of a rampaging sea monster named Cetus, who began to destroy the coastal settlements of her country. This monster could only be stopped by the sacrifice of her daughter, Andromeda. Reluctantly, Cassiopeia and Cepheus chained their tearful daughter to the rocky shore and left her to her fate. Fortunately, just as the monster was about to devour her, Andromeda was saved by the hero Perseus, who happened by just after he had destroyed the Gorgons. In fact, he was carrying Medusa's severed head in his bag, and after telling Andromeda to shield her eyes, he brandished Medusa's head in front of Cetus and the monster turned to stone. According to Ovid's version of this myth in his *Metamorphosis*, Perseus bargained with Andromeda's parents over her hand in marriage in return for destroying the sea monster, a bargain that they gladly honored. Many of the figures in this story have also had constellations named after them, particularly Andromeda's parents, who both have their own constellations.

Just off the inside curve of Andromeda is what looks to be a small, fuzzy star. Often misnamed as the Andromeda Nebula, it is the most distant object visible to the naked eye. It is, in fact, our sister galaxy, also named Andromeda, a colossal, spiral galaxy that is twice the size of our own and more than two million light-years away. The Milky Way galaxy and Andromeda are tilted, relative to each other, in near symmetry. If there were astronomers on a planet within Andromeda and they turned their telescopes our way, our galaxy would look very much as theirs does to us, though a little smaller. Curiously (because most galaxies are retreating from each other as the universe expands) Andromeda and the Milky Way are on a collision course, hurtling toward each other at a speed of three hundred thousand miles an hour. There's no need to worry though, at least not immediately—three hundred thousand miles an hour is an inter-stellar snail's pace—the collision won't take place for several billion years.

The Southern Cross, also known as Crux, is one of the few constellations that can't be seen from the Northern Hemisphere. It has been made famous mainly because of its appearance on the Australian flag, the only national flag that accurately depicts a constellation. The Southern Cross is truly a southern constellation: in order to see it you have to be south of the Tropic of Cancer, and even then it just peeks above the horizon in the month of May. Although it has garnered few mythological legends, it was an important constellation for the early European explorers of the southern seas because it pointed to the south celestial pole, which, unlike the north celestial pole, has no star to mark it. Since the Southern Cross lies within the Milky Way, it sometimes has a misty look, as if the hazy tropical nights it presides over somehow stretch up into the stars themselves.

The three bright stars of Orion's belt are usually the first star pattern that children learn to recognize after the Big Dipper. His form is fairly easy to imagine: a dark area above the belt marks his head, and the two bright stars just below the head are his shoulders. One of these stars, the one that marks his right shoulder, is the red supergiant star Betelgeuse. Two more stars beneath the three stars of his belt mark knees. From his belt hangs a sword, traced by three faint stars, the middle one of which looks a little fuzzy, because it is actually a nebula—the Orion nebula—a cloud of dust and gas within which the youngest stars in our galaxy are forming. Orion's cultural legacy stretches from a mention in Homer's *Odyssey* to the logo for the modern-day film company Orion Films, and it has inspired legends and myths for millennia.

Orion can be seen from anywhere on earth. The ancient Arabs knew Orion as *al Jauzah*, and they called the three bright stars of Orion's belt the "string of pearls." The Egyptians knew Orion as the resting place of their god of the underworld, Osiris, after he was murdered by Set. According to ancient Indian mythology, the constellation we call Orion was known as Prajapati, the lord of creatures and an incarnation of Brahma. Prajapati fathered twenty-seven daughters who became the wives of King Soma, the moon. But Prajapati was an incestuous father and one day he tried to catch one of his daughters, Rohini, to seduce her. A nearby star (known as Sirius) was outraged and pierced Prajapati with an arrow. The three bright stars of Orion's belt make up that arrow.

Curiously, Greco-Roman mythology also associated Orion with the moon, jealousy, and murder. In Greek mythology Orion was the son of Poseidon and a great hunter. According to one legend, Diana, goddess of the moon and the hunt, fell in love with Orion so desperately that she neglected her nightly transit across the sky. This disruption of the natural order of things worried the other gods, but none could convince Diana to take up her nocturnal duties. One day Diana's brother, Apollo (the sun), saw Orion swimming far out in the Mediterranean ocean. Using his light to obscure Orion's form Apollo then taunted his sister to hit the mysterious seaborne target with one of her arrows. She took the challenge and only later, when Orion's impaled body washed ashore, did she realize that she'd been tricked into murdering her lover. In her grief she placed his body in the sky and marked it with the bright stars we know today as the constellation of Orion. After Orion's death Diana became inconsolable and lost all interest in life, which is why the moon has been cold and lifeless ever since. Ursa Major, also known as the Big Dipper, is one of the most familiar constellations. Children are quick to see the shape of the Big Dipper because it is completely outlined by stars. In other words, there is a one-to-one relationship, like a connect-the-dots drawing, between the stars and the shape they make. This is also true of Ursa Minor, or the Little Dipper. "Seeing" most constellations, including Orion, is a bit of a stretch because you have to fill in gaps, like reconstructing an ancient vase in a museum where only a few shards provide clues to the rest of the shape. But Ursa Major provides the shape not only of a dipper, but also of its namesake. Although highly simplified, Ursa Major's stars actually do follow the outline of a bear, an abstract, simplified stick-figure bear, but a bear nonetheless.

The interpretation of Ursa Major as a bear has been acknowledged by many world cultures. The ancient Greeks called this constellation Arktos, or Bear—*Arktos* is the etymological root of the English word *Arctic*—because the constellation of the bear presided over the extreme north. According to Greek mythology Zeus became enamored of a beautiful young woman named Callisto to whom he appeared in the form of the goddess Diana in order to seduce and ravish her. When Hera, Zeus's wife, found them out, she turned Callisto into a bear. Zeus was bereft, and he placed Callisto, in her incarnation as a bear, into the sky along with her son, Ursa Minor. Two American Indian tribes, the Algonquian and the Housatonic, also saw the shape of a bear in these stars. According to the Housatonics, the three stars that form the handle of the Big Dipper were three hunters pursuing the great bear. Every autumn the hunters caught up to the bear and loosed their arrows at him, several of which pierced the bear, causing it to bleed. The Housatonics believed that Ursa Major's blood, falling to the earth at night, was what colored the leaves red in the fall.

In England the stars of the Big Dipper have often been seen as a wagon or a plow, and in Anglo-Saxon mythology the constellation was associated with King Arthur. The fact that these stars were seen as a celestial Camelot revolving around a central point is reputed to have given rise to the concept of King Arthur's Round Table. The British also used the counterclockwise rotation of the Dipper stars around Polaris as a way of telling time at night, for the stars revolve like the hands of a clock.

Another of the best-known features of the night sky is the Milky Way, which meanders through the constellations like an irregular banner of misty light. It is a hypnotic, thrilling sight, even if you don't conceptualize it for what it really is—a view of our own galaxy, composed of billions of stars. Because we are looking at the disc of our galaxy from the side, we see only a line, and not the true, circular shape of the Milky Way. There is a profound sense of mystery to this spectacle of age and size, of immensity beyond conception. Here, in our place on an outer arm of our spiral galaxy, we look toward the core of the Milky Way, though we cannot see it because it is obscured by dust and gas. Our sun, and all the other stars in the Milky Way, orbit once around the center of the galaxy every 220 million years. At the core of the Milky Way, around which all the stars of the galaxy revolve, there is believed to be a massive black hole, or perhaps several massive black holes orbiting each other. Black holes are the ultimate cosmic objects and are the night of night in outer space; even light cannot escape them and is extinguished forever in their depths.

Massive black holes, like the ones that are thought to form the nucleus of our galaxy, swallow stars and grow larger. According to Stephen Hawking, the world-famous astrophysicist and professor at Cambridge University, when a star is destroyed in a black hole, all the information that was contained in that star—its planets, the histories of any civilizations that existed on those planets, the structure of each and every atom that made them up—all of it disappears, and is erased for all time, because even black holes don't last forever. Hawking believes that they eventually "evaporate," leaving nothing to show they ever existed. Only the light that the star emitted before it fell into the black hole remains behind—only the twinkle of yet another, anonymous star is left to speed on its finite, one-way journey to the end of time.

Falling Stars

One mid August a few years ago, when I was camping at a National Park on Lake Huron, I saw a notice at the interpretive center inviting campers to participate in an annual Perseid meteor shower observation project. It was an evening event that was being organized by a local group of amateur astronomers. I signed up, and learned that along with other volunteers and seasoned meteor observers, I would be recording the paths of meteors on sky maps. All I needed to bring was a reclining lawn chair, the type you use to sunbathe at the beach, a blanket, and a pair of binoculars.

I arrived an hour after dark, just at the beginning of astronomical twilight. The parking lot had about ten cars in it with some people still unloading chairs and blankets from their trunks. I gathered up my own lawn chair and blanket and walked over to the lawn in front of the nature center toward the observation area. There were already half a dozen lawn chairs arranged in a neat circle. Two meteor-watchers were lying back in their chairs, binoculars pointing up at the sky, and a couple of kids were horsing around with flashlights. Gleaming whitely in the darkness at the perimeter of the circle were several large, reflecting telescopes that had been set up by their owners. You could see the tubes that held the big reflecting mirrors pointing at various parts of the sky. A tall woman named Andrea, with long curly hair pulled back into a ponytail, was in charge. "You'll be glad you came," she said. "They're predicting a good display tonight." That night, August 12, was the height of the Perseids meteor shower, a display of meteors that occurs every year at the same time. This is because the earth passes through a "dust belt" at the same point in its orbit around the sun each year. Andrea then handed out sky charts that had a few "marker stars" superimposed over a grid. Whenever we saw a meteor, she told us, we should mark its length and direction on the chart.

I returned with everyone else to the ring of chairs, where we lay down with out heads toward the center of the ring, holding our sky charts and pens poised. Andrea stood in the center of the ring and explained that most of the meteors we saw were probably no larger than grains of sand, but that we'd never know it from the size and length of the flare they caused as they burned in the atmosphere. Almost as soon as she finished, we all saw a brilliant falling star that was greeted by a few gasps. It was very bright and streaked across the whole sky, leaving a faint trail that lasted for half a minute afterward. "How large was that one?" I asked Andrea in the darkness, presuming we'd seen something unusually large. "Maybe the size of a pea," she replied, "but a pea," she went on, "that's moving at ninety thousand miles per hour makes quite a flare."

The Message from the Stars

The treasures hidden in the heavens are so rich that the human mind shall never be lacking in fresh nourishment.

JOHANNES KEPLER

Astronomers have been gazing at stars for millennia. The Mayans were expert stargazers, and they calculated the orbits of the planets and the phases of the moon with such precision that their calendars are still accurate today, not just in predicting lunar and solar eclipses but also in predicting the conjunctions of the planets. The Egyptians were famous stargazers as well, and many architectural aspects of their pyramids lined up with celestial objects. Both the Mayans and Egyptians made their observations from their temples, which doubled as observatories and places for the performance of religious rituals. The first observatory dedicated solely to astronomy was constructed by the Egyptian king Ptolemy I in Alexandria in 300 B.C. Coincidentally, more than five hundred years later, the Greco-Egyptian mathematician Ptolemy, though not a descendant of Ptolemy I, became the greatest astronomer of the second and third centuries. He mapped 1,022 stars into forty-eight constellations, and his map of the stars, the *Almagest*, was in use until the time of Copernicus. In the ninth and tenth centuries, during the great mathematical and scientific flowering of the Islamic empire, astronomical observatories were built at Baghdad and Damascus. But the real revolution in astronomy had to await the European Renaissance.

The first of the great astronomers, Nicolaus Copernicus, was born in 1473 in Torun, Poland. He studied astronomy at the University of Krakow and then moved to Italy where he continued his studies at the world's two oldest universities, Padua and Bologna. After attending these universities, Copernicus moved to Rome where he lectured on mathematics and astronomy. The years that he spent in Rome, from 1500 to 1512, were extraordinarily fertile ones. Northern Italy was the epicenter of the High Renaissance, and a new rationalism was flowering in the superstitious rubble of the Dark Ages. But it was when Copernicus read ancient Greek works on planetary motion, particularly the writings of Aristrarchus, that he began to work on his heliocentric theory of the solar system.

He moved back to northern Europe in 1512, though he continued to write and to visit his friends and colleagues in Rome. Eventually he became the canon of the cathedral of Frauenberg in Poland, and on every clear night he would mount one of the turrets on the cathedral's wall and make meticulous observations of the stars and planets. His astronomy was conducted entirely by "eyeball," as the telescope had not yet been invented. The only instruments Copernicus used were a triquetrum, a sundial, and an astrolabe. Even these were sometimes useless; the Frauenberg cathedral was so close to the Baltic Sea that his vision was often obscured by fog.

As Copernicus compiled his observations, he became convinced that the earth and all the planets orbited the sun. It was to his friends in Rome that he first explained this revolutionary idea and they encouraged him to publish his observations. His contentious theories were eventually printed in 1543, as he was dying. His book, called *De revolutionibus orbium coelestium*, was dedicated to Pope Paul III, which was rather ironic considering the fact that Copernicus's theory was thought blasphemous. The resurrection of Greek and Roman culture and science that occurred during the Renaissance had become a great challenge to the church. And the most sacrilegious information seemed to be emerging from the night sky, as a handful of astronomers began to cast their mathematical nets into the stars. The picture of the universe that began to emerge was to prove almost fatal for the Roman Catholic Church.

Outside of Italy the climate of the Renaissance was not quite as frostbitten by the chill of the Inquisition, and astronomers were sometimes able to publish their findings without fear of immediate reprisal by the Vatican. The two most prominent of these were Johannes Kepler and Tycho Brahe. Brahe was a Danish astronomer who had dominated astronomical science in the sixteenth century. He died just before the invention of the telescope, but nevertheless made meticulous observations of the stars from atop the observatory that surmounted a castle he built for himself on the island of Ven, which he called Uranienborg. He agreed with Copernicus that the planets orbited the sun, but he excluded the earth and moon from this system, declaring instead that the sun orbited the earth and moon. The much younger German astronomer Johannes Kepler was Brahe's assistant at the Prague observatory in Czechoslovakia at the turn of the seventeenth century. Kepler continued Brahe's work after he died and, wisely, continued to maintain Brahe's exclusion of the earth from the Copernican heliocentric theory, stating that the sun revolved around the earth and moon. This was a position that barely satisfied the jumpy Vatican cosmologists, though according to pontifical law there wasn't enough heresy in his theory to light an Inquisitional bonfire under Kepler's feet.

The explorations of space end on a note of uncertainty . . . we measure shadows . . . we search among ghostly errors of measurement.

EDWIN HUBBLE

Eight years prior to the turn of the century, in 1592, at the height of
the Baroque period, the astronomer who would perfect Copernicus's
system and in the process make himself both famous and a marked
man was appointed professor of mathematics at the University of
Padua in Italy. He was an unconventional lecturer and privately
resented having to interrupt his own research to teach students.
While teaching in Pisa he had discovered several laws of physics that
contradicted Aristotle, including the constant velocity of falling
bodies of different weights. According to legend, he publicly proved
his theory by dropping two balls of unequal weight from the top of
the Leaning Tower of Pisa. His name was Galileo, and in the last
decade of the sixteenth century he corresponded regularly with
Kepler. By 1598 he stated in one of these letters that he believed
in the heliocentric theories that Copernicus first proposed in 1543—
namely, that the earth revolves around the sun and not vice versa. He
knew this was a heretical statement, and to say so in writing, even if to
a like-minded colleague, was risky business. But, luckily for Galileo,
he was soon to have observational ammunition for his claims. Galileo
was perhaps the most fortunate of the great astronomers in having
been born late enough to take advantage of a technological innovation
that would revolutionize the study of the stars.

In 1578 William Bourne published a book called *Inventions or
Devices*, in which he described a remarkable apparatus he had constructed: "For to see any small thing a great distance from you, it
requireith the ayde of two glasses," and with such an arrangement you
could "see a man four or five miles from you." By 1608 his telescope
was appearing all over Europe and in the same year a patent
application for the device by Hans Lippersheim, a Dutch lens maker
from Middleburg, was turned down. The cat was out of the bag and
though somewhat of a novelty item, like the camera obscura that
preceded it, its reputation spread quickly across Europe.

When Galileo heard of this new invention, he immediately realized its true importance. In the summer of 1609 he built his own telescope and used it to secure himself a pension from the Venetian senate after he had proved its value to a group of senators by inviting them to look at a distant building through its eyepiece. Later that year he built an even better telescope, one with a magnification power of twenty times. On a warm night in November 1609, he set up his new telescope in his garden at Padua and aimed it at the stars. He looked at the moon and could distinguish mountains, he looked at Jupiter and discerned what appeared to be small stars circling it. Every clear night that winter, Galileo could be seen staring through the eyepiece of his device into the night sky. He kept a candle lit beside him so that he could take notes and make his meticulous drawings of the moon and planets. The very next year Galileo published his first astronomical observations in a short book called *Message from the Stars* (*Siderius Nuncius*). It created an explosion of controversy. In it, Galileo proved that the Milky Way was made of millions of small stars. He also reported that he had seen mountains on the moon and that he had observed four moons orbiting Jupiter. The universe would never be the same. After he began to calculate distances and orbits, the universe became even larger, and more impersonal than it had ever been.

Galileo, always with an eye on career opportunities, named the four moons of Jupiter the "Medicean stars," after the Grand Duke of Tuscany, Cosimo de' Medici. His nominative homage paid off and before long he was appointed as the court mathematician and philosopher of the Medicean court. In Florence (which he detested) Galileo continued his research and observations. Over a series of very clear nights in 1613, using an improved telescope that he had just made, Galileo observed that Venus had phases, just like the moon. Every night, it seemed, he was seeing deeper and deeper into the secrets of the universe, discovering more strange truths about the planets and the stars. It became increasingly obvious to Galileo that

Copernicus was right and he expounded this opinion to anyone who would listen. Galileo was not a retiring or shy man; he seemed to live by William Blake's maxim from "The Proverbs of Hell": "Speak your mind and a fool will avoid you," and when he believed in something, he presumed everyone else would eventually come round to his viewpoint if they saw the evidence. But others found Galileo's temperament not only a bit abrasive, but also blasphemous.

It remained a dangerous time to be in opposition to the Vatican. In 1616 the dreaded Inquisition secretly warned Galileo, even though he was under the protection of the Grand Duke of Tuscany, to be a little less strident about his heretical views, at least in public. This infuriated Galileo. If these doubters simply looked through his telescope then they could see the evidence right there, in the stars! But they saw a much different universe regardless of the optical evidence. Galileo decided he had to win his opponents over with overwhelming arguments. During the next few years he worked on a manuscript that he hoped would make everything clear. Published in 1632 and titled the *Dialogue of Two Systems*, Galileo's manuscript was a masterpiece of rhetorical insinuation. It consisted of a dialogue between two men—an orthodox skeptic versus a scientific pro-phet—who weighed the pros and cons of whether the sun orbited the earth or the earth orbited the sun. In *Dialogue* Galileo used his characterization of the skeptic to paint an unflattering portrait of the Roman Catholic Church's position. Of course *Dialogue* was a public-relations disaster, and Galileo's hopes for rationalism on behalf of the Vatican were dashed.

Shortly after publication of *Dialogue*, an ailing Galileo was sum-moned to Rome where he was found to be in heretical breach of the church's tenets. It is said he was forced to kneel and recant his heresy before the Inquisition. Considering the times, and the murderous reputation of the Inquisition, Galileo got off lightly. He was con-demned to house arrest for life and forbidden to publish. Neither of these punishments proved to be onerous; his house-arrest sentence was carried out at his country villa in Arcetri, near Florence, and his succeeding books were smuggled out of Italy to be published in Holland. Perhaps the Inquisition had softened in the last few decades, perhaps it had rightfully begun to have second thoughts about destroying the most brilliant minds of Italy. After all, in 1600 it had burned the genius Giordano Bruno at the stake for suggesting, among other things, that space was infinite and that our solar system was but a small part of a much larger universe!

Ultimately Galileo was lucky. He lived to enjoy a quiet, pastoral existence at his country villa, where he had an extensive garden. In one sense his house arrest was a boon as it allowed him to work in peace and to maintain his garden, the two things he loved the most. And every night, when the stars began appearing in the opaline Italian evening, he must have felt a celestial vindication. After all, wasn't the only proof he would ever need right there in the dark sky, and if knowledge and rational thought prevailed, then wouldn't he be ultimately exonerated? The floodgate of knowledge, once opened, could never be closed, and he knew in his heart his name would be immortal.

In questions of science, the authority of a thousand is not worth the humble reasoning of a single individual.

GALILEO

Moonstruck

All I want is the moon, Helicon. I know in advance what will kill me. I have not yet exhausted all that can make me live. That is why I want the moon.

ALBERT CAMUS

Wolves howl at it, lovers swoon under it, and the oceans are pulled by it. It has inspired more poems and songs and myths than any other celestial body including the sun. We've landed astronauts there and brought back pieces of it to earth, yet the moon remains as mysterious and enchanting as ever. As Joseph Conrad wrote: "There is something haunting in the light of the moon; it has all the dispassionateness of a disembodied soul, and something of its inconceivable mystery." The moon has been called "the queen of the night" and it epitomizes darkness more than any other celestial body. Its phases are a part of the rhythm of our lives and it has been a celestial fixture for our entire history. A crescent moon is the international icon of evening and its presence on television weather forecasts, motel signs, and medicine containers signifies night.

From children's stories to Shakespeare, from cartoons to Academy Award—winning films, from cave art to Renaissance masterpieces, the moon has been serenaded, confessed to, and longed for. It is deeply embedded in our language. If you are a bit dazed, people say you are "moonstruck." If you have a wide, featureless face and a dreamy disposition, you are called "moony." If you are a born fool, you are a "mooncalf," and if you are ecstatic about something, they say you are "over the moon." To "cry for the moon" is to want something you cannot have, yet to "promise the moon" is a wildly extravagant proposition meant to win someone over. Hill folk make "moonshine" and "honeymooners" "moon about" in lovestruck idleness.

The moon is ancient and timeless. As Thomas Hardy wrote: "Shut out that stealing moon/She wears too much the guise she wore/Before our lutes were strewn/ With years-deep dust, and names we read/On a white stone were hewn." Perhaps, above all, the moon is romantic. It illuminates the chaste kisses of virgin lovers and the lustful unions of seasoned voluptuaries. It is also erotic, as Christopher Fry declaimed: "The moon is nothing/But a circumambulating aphrodisiac/Divinely subsidized to provoke the world/Into a rising birth-rate."

Lunar phases provided the first monthly calendar for humans, and several religions, notably Judaism and Islam, still base the timing of their observances on the lunar cycle. In the Lascaux caves in south-western France, alongside an exquisitely rendered horse, is a series of dots that are thought to represent the moon's cycle. If they are indeed a record of a lunar month, then this is among the earliest calendars, well over fifteen thousand years old. The first confirmed calendrical representations of the moon are on Babylonian clay tablets that date from 750 B.C., but the various phases of the moon represent not only a nocturnal, sky-born calendar: they also have many mythological and literary allusions, particularly the crescent moon. Many statues and paintings of the Roman goddess Diana showed her standing on a crescent moon, a reference to her role as a lunar deity. In the Hebrew Bible the crescent points of the new moon were called horns, and it was common for the Israelites to say that great or wise persons had a light around their heads like the horns of the new moon. It is thought that an overly literal translation of this saying is the mistaken precedent on which Michelangelo based the horns of wisdom he placed on his sculpture of Moses.

The Hare in the Moon

What people see "in" the moon is as various as the cultures they come from, and the lunar myths they were raised on. The Maori of the South Pacific see a woman with a bucket in the moon. According to their legends her name is Rona and, as daughter of the sea god Tangaroa, she controlled the tides. Once she stumbled in the dark and cursed the moon for not being bright enough to illuminate her path properly. The moon, in revenge, imprisoned her on its surface, which is where you can see her today.

The Hindus see a hare in the moon, which is also why hares are regarded as incarnations of the god Soma, the Hindu moon god. He drives his chariot, pulled by white horses, across the sky and radiates bliss upon the world. A divine elixir that bestowed eternal youth and happiness was also named Soma, after the moon god, because it was thought that the lunar surface was covered by oceans of it. Perhaps that is how the Hindus explained the moon's intoxicating effect on both man and beast.

The sacrificial violence of Mesoamerican civilizations is well represented in the Aztec lunar myth. The Aztecs called the moon Coyolxauhqui, which translates as "Golden Bells," because they saw a princess's face with golden bells in the moon. As in many other mythologies, Coyolxauhqui was the sister of the sun god, Huitzilo-pochtli. According to legend, Coyolxauhqui was killed by Huitzi-lopochtli after he discovered that she had instigated the murder of their mother, Coatlicue. Huitzilopochtli also killed Coyolxauhqui's four hundred brothers and sisters, although, after he killed his sister, he was so overcome by remorse he cut off Coyolxauhqui's head and tossed it into the heavens, where it glowed with the golden hue of her bells forever afterward.

The gender relationship in the Japanese Shinto ("the way of the gods") moon myth is the reverse of that of the Aztec myth. Here the moon is male and the sun is female. Shinto legend relates how both the sun and the moon gods were born from the eyes of the original being, Izanagi. Tsuki-Yomi, the moon god, was born from Izanagi's right eye, while his sister, Amaterasu, the sun goddess, was born from the Izanagi's left eye. They lived together as siblings until Tsuki-Yomi publicly humiliated them both by murdering another god, the goddess of food, for serving him a disgusting meal. Amaterasu was outraged when she heard this and since then they have always lived apart, which is why you rarely see them in the sky together.

The Blackfoot Indians of North America see a woman in the moon. In their cosmology the sun and moon were husband and wife and they had seven sons (the stars of the Big Dipper.) The sun god made the world, though after finishing the creation he felt he had made a mistake by creating snakes and he decided to destroy them all. In the end, though, he couldn't bring himself to kill a female snake that was pregnant, so he spared her. One of her descendants became Snakeman, a powerful being who decided to avenge the destruction of his ancestors by making the sun suffer. In human form, he seduced the moon and they began an adulterous affair. One day the sun discovered them and killed Snakeman. But the moon, who had fallen deeply in love with Snakeman, became enraged at the sun and chased after him and their sons. Fearing for his own safety and the safety of his sons, the sun fled westward, chased by the moon. Finally, in desperation, he flew up into the sky to try and avoid her, but she rose right up after him. Using his powers one last time, the sun divided night from day to gain a resting period from the pursuit. Throughout the day he and his sons could rest, but at night, the chase was on again. The process is unending, and even today, according to the Blackfeet, the moon pursues the sun and their seven sons across the skies.

The Chukchi of Siberia have a legend called the "Reindeer Maid," which is very similar to the myth of Selene and Endymion, except that, like the Shinto legend of Japan, the Siberian moon god was male. He fell in love with a young reindeer shepherd who played beautiful songs on her flute every night. One night he came down to earth and tried to kidnap her but she tricked him and trapped him. Finally after much pleading she released him, making him promise to never come down to earth again, a vow he keeps to this day, though his loneliness sometimes makes his glow weaken.

The Chinese see a toad as well as a woman in the moon, while the Japanese see a hare, like the Hindus. The Japanese also have a poignant legend about an unrequited love between the daughter of the moon and a prince. This legend accounts for the origin of fireflies. The moon princess came to live on earth, blessing a mortal family, and falling in love as a young woman with a prince. Unfortunately she could only inhabit the earth until she was twenty years of age, and on the night she returned to the moon, she shed many tears over her lost love for the prince. As her magical tears approached the ground, they began to glow bright green and turned into fireflies. This theme of lunar tears was also the basis of an ancient Peruvian Indian account of the origin of the precious metal silver. They believed that silver nuggets were the moon's tears, given to humans as a legacy of the moon's generosity and loneliness.

Queen of the Night—Moon Lore

Enchantress of the stormy seas,
Priestess of Night's high mysteries.
SARAH HELEN WHITMAN

In the Northern Hemisphere every month seems to have its own moon. In February there is the snow moon and in May the planting moon. The honey moon presides over June nights and is considered both the lover's moon and the moon of marriage. Summer's dog days bring with them the grain moon of August and in late September comes the harvest moon. Sometimes the harvest moon falls in October, in which case it doubles as the hunter's moon. In China the harvest moon is recognized by a festival whose origin is in a romantic legend. According to the legend, in ancient times there was a beautiful woman called Chang Er who was tormented by her husband. She chanced upon a magic potion that transformed her into a fairy, whereupon she flew to the moon. It is said that when the harvest moon is at its brightest, it is possible to see Chang Er there.

A harvest moon!
And on the mats—
Shadows of pine boughs.
KIKAKU

The full moon closest to the autumn equinox is called the harvest moon because, before the advent of floodlights on tractors and harvesting machines, farmers were able to take advantage of its constant light on clear nights to continue harvesting their crops well past sundown. But is the harvest moon really that different from other moons? Surprisingly, it is. In the Northern Hemisphere the harvest moon rises more or less at the same time every night, instead of much later each successive evening as it usually does the rest of the year. This is because the moon's orbit is closely aligned with the ecliptic plane, the flat disc formed by the orbits of the planets in our solar system as they circle the sun. For most of the year the ecliptic plane intersects the horizon at a steep angle, so that on each successive night the moon is farther below the horizon because the moon moves eastward relative to the sun. On these nights, which make up most nights of the year, the moon rises fifty minutes later each successive evening. That relationship between the ecliptic plane and the earth's horizon changes in September, during the equinox, because at this point of earth's orbit around the sun, the angle of the ecliptic plane, and therefore also the moon's orbit, is much more shallow relative to the horizon. This means that the first full moon closest to the September equinox is not as far below the horizon on succeeding nights, so that it rises only thirty minutes later each successive evening rather than fifty minutes later.

But the moon has other tricks up her sleeve. Due to a cosmic coincidence, the diameter of the moon viewed from earth is almost exactly the same as the diameter of the sun (roughly the same size as a

Canadian or American dime held at arm's length). This means that during total eclipses the moon completely obscures the sun and an eerie dusk descends in the middle of the day. Birds stop calling and the sky takes on a purple, twilit hue. No wonder the ancients attributed so much power to the moon—she was capable of turning day into night.

During a lunar eclipse the earth's shadow turns the moon a copper color, though according to legend it can also turn blue. We say "once in a blue moon," which means very rarely, and by definition, "blue moons" do occur quite infrequently—once every two and a half years or so. Though many people believe that a blue moon is a moon that appears blue in color, the term actually refers to a second full moon falling in the same calendar month. When, during the first sliver of the new moon, you can see the rest of the moon's disc if it is illuminated by earthglow (sunlight reflected from the earth to the moon), it is called "the old moon in the new moon's arms." Sometimes there are haloes around the moon, and sometimes, on earth, there are even lunar rainbows, like the one that can be viewed at Victoria Falls. During the peak flow of the Zambezi River, running along the border between Zimbabwe and Zambia, an incredible 17.6 million cubic feet of water thunders over the 5,604-foot-wide rim of the falls per minute. The roar of the water as it plunges 420 feet into the gorge below can be heard twenty-five miles away, and the plume of mist from the falls can be seen from a distance of fifty miles on a clear day. But Victoria Falls' most sublime sight occurs at night during the full moon. In bright moonlight a pallid, nocturnal rainbow forms an arc across the mist-filled gorge. The lunar rainbow of Victoria Falls has been described as eerily monochromatic, though some observers have claimed to see pale violet and blue bands in its spectrum.

Lunacy

It is the very error of the moon,
She comes more near the earth than she was wont,
And makes men mad.

WILLIAM SHAKESPEARE

There is a long, folkloric tradition of attributing to the moon a special influence over human behavior. By Shakespeare's time it was common knowledge that the moon affected certain people unduly, and that during the full moon there was an increase in unusual events. "Moon madness" was taken seriously enough at the time that the term *lunacy* was coined, taking as its root the Latin word for the moon, *luna*. Paracelsus, the famous sixteenth-century physician, diagnosed many cases of lunacy in his day, and by 1842 a diagnosis of lunacy was officially recognized by British law under the terms of the Lunacy Act. The law declared that someone was a lunatic if they had "lucid intervals during the first two phases of the Moon" but were "afflicted" with "fatuity" in the "period following after a full moon."

Now, supposedly, we know better, at least as far as the moon's influence over behavior is concerned, although there do remain a few traces of the moon's influence over our physical selves. It has been more than four hundred million years since our ancestors crawled out of the Devonian oceans, yet the lunar rhythms that dominated their oceanic existence still linger in our bodies. Menstruation (derived from the Latin *mensis*, or "lunar month") follows the 29.5-day lunar calendar, as does the average time for human gestation, 266 days, which is equivalent to nine lunar months. Women, it seems, are linked to the moon. The word *month* itself is derived from the word *moon*. But, other than these somewhat tenuous connections, there is no scientific evidence to support the claim that the moon has any influence over our lives. Despite this evidence to the contrary, beliefs about lunar influence have not waned, indeed, in the twentieth century they have probably expanded, rising on a tide of pseudoscience. In fact, the myths of lunar influence on various human phenomena became so persistent and prevalent over the last few decades that three psychologists, I. W. Kelly, Roger Culver, and James Rotton, decided to publish their statistical analyses simply to put to rest the claptrap that had proliferated during the twentieth century.

In their article, first published in 1985–86, the authors assembled the statistical data from more than one hundred studies on lunar effects and found that there was no significant correlation between a full moon, or any phase of the moon for that matter, over any of the following phenomena: emergency-room admissions, the homicide rate, major disasters, psychiatric admissions, traffic accidents, crisis calls to police or fire stations, domestic violence, alcoholism, births of babies, suicides, casino payout rates, assassinations, kidnappings, violence in prisons, agitated behavior by nursing-home residents, assaults, gunshot wounds, stabbings, lycanthropy, vampirism, sleepwalking, or epilepsy.

So, with a complete lack of statistical evidence for any of these effects, why do we continue to believe that the moon affects us? Kelly, Culver, and Rotton think that we are conditioned by media, folkloric traditions, superstition, and, more recently, urban myths. They insist that people questioned about their claims of lunar influence closely cannot muster much evidence for their conviction. Perhaps people's confidence in the moon's power does come partially from societal conditioning, but also from our thrall of what Walt Whitman called "ghastly, phantom moon;/Immense and silent moon." Even in our rational age the moon continues to exert a mesmeric influence.

The Honey Moon

A savage place! as holy and enchanted
As e'er beneath a waning moon was haunted
By woman wailing for her demon-lover.
SAMUEL TAYLOR COLERIDGE

The moon is the nocturnal deity of love, and its romantic influence is constantly reinforced. As the song "That's Amore" had it, when the moon looks like a deluxe pizza you're infatuated. But the moon isn't just a barometer of enamoration, it is a kind of celestial mascot for lovers and honeymooners. When I was growing up in the late fifties, I remember watching the television program *The Honeymooners* with Jackie Gleason, where the credits were superscripted over a big moon that, at the end of the introduction, transformed into Jackie Gleason's face in a kind of subversive personification. I think that was the first time I realized our cultural association between the moon and romance.

It was when I began watching films that the connection was really solidified. Hollywood has depicted the amorous moon in too many films to mention. The cinematic scenes almost always feature an elevated balcony or a terrace overlooking the sea, on which the lovers kiss by the light of a full moon. In Frank Capra's *It's a Wonderful Life* (1946) there's a very romantic lunar scene where James Stewart, on his first date with his future wife, Donna Reed, stands below a beautiful full moon that he offers to lasso for her. Years later, in a 2003 comedy starring Jim Carrey called *Bruce Almighty* this scene was revisited, though with a literal twist, as "Bruce" actually lassoes the moon and tugs it toward the earth to impress his fiancée. Although historically there has always been a folkloric recognition of a lunar influence over love, it was the Romantic poets who first popularized the link between the moon and romance in the early nineteenth century.

In poems like Percy Bysshe Shelley's "Queen Mab," the moon and night are a prerequisite proscenium for love: "Heaven's ebon vault / Studded with stars unutterably bright / Through which the moon's unclouded grandeur rolls, / Seems like a canopy which love has spread." John Keats penned a poetic tribute to the moon in which he wrote: "When soul meets soul on lovers' lips, / high hearts are calm, and brightest eyes are dull; / So when thy shadow falls on me, / Then I am mute and still by thee / Covered; of thy love, Orb most beautiful, / Full, oh, too full!"

The Moon's Origin

This is the light of the mind, cold and planetary.
SYLVIA PLATH

There are many theories concerning the origin of the moon. The leading contender among current scientific explanations is the "giant impact theory," which has steadily gained credibility since the mid 1980s. It suggests that our moon was created out of debris from a collision between earth and a Mars-sized object approximately 4.5 billion years ago, just after the formation of the earth. The ring of debris created by this collision orbited the earth and condensed into our moon. According to one researcher, Japanese astrophysicist Eiichiro Kokubo, the moon formed into its present, spherical shape in just one month. The moon is largely made up of whatever it was that hit the earth.

The leading theory prior to the giant impact theory proposed that the moon was formed by accretion out of the same cloud of primordial dust as the earth. Still another theory, though it is now discredited, suggested that the moon was a stray planet that was captured by earth's gravity. The fourth major theory of the moon's origin is one that has also fallen into disfavor, although it is my personal favorite because it is the most grandiose and poetic. According to this theory the moon was composed of material thrown off by the earth when its rate of spin was increased by the migration of heavy elements into the earth's core during its early formation. The Pacific Ocean fills the hollow left when the earth gave birth to the moon.

When I read this theory, I formed a literal fantasy of the moon's birth that centered on the spectacle of the moon rising, fully formed, out of the Pacific Ocean. The first moonrise. A moonrise where the moon just didn't appear to be rising out of the ocean, but actually was, with millions of tons of seawater and moondust pouring off its sides as it ascended on an ocean column into the stars. And within this umbilicus of seawater swam rainbow-hued Pacific fish, sharks, whales, and phosphorescent deep-sea creatures glowing inside the pillar of ocean connecting the moon to the earth. (And weeks later, in my fantasy, after the Pacific umbilicus had broken and the few remaining pools of ocean water had dried up on the moon's surface, these tropical fish would lie flopping in the moondust thousands of miles from their native seas.)

Of course now that we've been to the moon, we've found no dessicated fish there, nor moonmen, nor any evidence of life whatsoever. But science hasn't taken all the poetry from the moon. When the *Apollo* astronauts visited its rock-strewn surface, the moon revealed a handful of secrets more subtle than any we had ever dreamed of.

The Apollo Moon Program

Here men from the planet Earth first set foot upon the moon, July 1969 A.D. We came in peace for all mankind.

PLAQUE LEFT ON THE MOON BY THE *APOLLO 11* ASTRONAUTS

The first manned exploration of the moon was one of the most heroic adventures we've ever undertaken. Nothing, in terms of earth exploration, compares to it—not in distance, not in cost, not in scale. But before we actually landed on the moon, NASA flew some practice missions to test the Apollo program's moon vehicles. The penultimate mission, the one just before the first manned landing, was called *Apollo 10*. Its mission was to test the lunar module docking procedure and—for the first time ever—to leave earth's gravity, fly to the moon without landing, and come back to earth.

Due to the phase of the moon during the mission, the *Apollo 10* astronauts flew silently toward a completely dark moon unmarked by even the slimmest crescent of light. Surprisingly there was no earthshine either, so it must have been disconcerting for them to be careening toward a moon that was not even a black shape in the sky, but was, instead, just a hole in the night, a dark place where there were no stars. As they approached the moon, the "hole" got larger and larger until it filled the entire view. It must have appeared to the astronauts that they were sailing into an eerie abyss of night inside of night itself. It was only after they went into orbit around the moon that they passed into the sunlight and saw the cratered lunar landscape spread out below them, less than twenty miles away.

It wasn't until the *Apollo 11* mission, though, that humans first walked on the moon and the rest of us, via television, got our first good look at something we had been gazing at, as a species, for thousands of years. It was an extraordinary moment to finally see the surface of the moon, to look at the rocks and mountains and moondust. To view, up close, the rocky surface that gives us moonlight, to see the austere, gray desert. Buzz Aldrin, the second astronaut to set foot on the moon, took in the lunar panorama for the first time and exclaimed, rather poetically, "Beautiful! Beautiful! Magnificent desolation." Other astronauts, like James Lovell, were more prosaic in their descriptions. Lovell said, "The moon is essentially gray, no color.

It looks like plaster of paris, like dirty beach sand with lots of footprints in it." As it turned out, the lunar soil had some unusual properties. Throughout the Apollo missions moondust got into everything, like a fine, sooty grit, and it proved almost impossible to wash off. Astronauts had it under their fingernails for weeks after they returned to earth.

Even though the *Apollo 13* mission deservedly got the lion's share of media attention, some of the quirkiest moments of the entire Apollo program came with the *Apollo 14* mission, when Alan Shepard and Ed Mitchell manned the moon base. Shepard recalled an incident after they returned from a day spent on an EVA (extra-vehicular activity). After they repressurized the lunar module, they ate dinner and then lay down to sleep. Shepard, the commander, was in the top bunk. He was having trouble falling asleep because the metal collar of his space suit was cutting into his neck. Just as he was drifting off, he heard a big "clang" noise that started his heart racing. "Ed, you hear that?" he whispered to Mitchell in the lower bunk. "Yeah, I heard it." Then they both broke into laughter when they realized how crazy it was for them to be whispering—what were they whispering for? Who could be out there trying to break in?

Perhaps if the *Apollo 14* astronauts had stayed long enough to endure the fourteen-day-long lunar night, they might have been even more jumpy, though the 400 degree Celsius drop in temperature might have made their lander a little too cold for sleeping. But even during the "day" that they explored, it still looked like night, because even when the sun is shining on the moon, you can see stars in the lunar sky. You can also see the earth. According to all of the astronauts their most poignant sighting was of the earth itself, shrunk to the size of a large marble, floating like a sapphire in the immensity of the night sky above the desolate surface of the moon. Certainly when we first saw pictures of our planet from space, it was an existential event for us, but to see the earth from that distance, floating like a cloud-swirled cerulean yolk above the cruel, empty beauty of the moon, was also to realize that our earth *must* be unique, an exception to the cosmic rule. It is likely, according to the most recent cosmological theories, that the vast majority of the billions upon billions of planets and moons in our universe are as barren as the moon and that earth is indeed a toehold of life in the infinite coldness of space.

Comets, Northern Lights, and Noctilucent Clouds

The only other celestial phenomena that rival the moon for eerieness are comets, even though their "heads" are made up of frozen hydrogen slush and bits of rock, and their "tails" are merely the gaseous vapor that is burned off by the heat of the sun as they approach the inner solar system. Comets orbit the sun like the planets, but their orbits often take hundreds of years. Perhaps the most famous is Halley's comet, named after British astronomer Edmund Halley, who in 1682 was the first to predict the return of a comet. His comet did indeed return in 1759, 1835, 1910, and 1986, though its most recent appearance did not have any of the drama of previous visits, where spectacular meteor showers accompanied the passage of the earth through the comet's tail. Some apocalyptic theorists continue to believe that disease comes from comets' tails in the form of interstellar microbes and viruses that rain down on earth and cause epidemics. They proposed this theory after a strange correlation was observed for several plagues that seemed to follow close on the heels of each visit by Halley's comet. Others think that perhaps life itself was seeded by biological material contained in comets, for comets come from the Oort cloud at the very edge of our solar system, and it is possible that some of the material in them could come from beyond the original cloud of dust and gas that our solar system formed from. Perhaps tiny spores of life do hitch interstellar rides on the backs of comets, seeding life across the cosmos.

Whatever you believe about comets, though, they do seem innately portentous. They look like a pale, cosmic exclamation mark or the glowing head of an evanescent, ghostly dragon. They also move unlike any other celestial object. They rise and set with the stars and moon, and yet every night they move relative to the stars, and much faster relative to the planets. There is something fascinating yet unearthly and menacing about a comet, as if it were a cold, phosphorescent angel of calamity—it hangs in the sky like a beautiful jinx. A comet is like a shard of congealed moonlight or willful clot of nebula that has gone wandering. Yet when a comet strikes, like the Schumacher-Levy 9 comet that hit Jupiter in 1994, its impact is far from nebulous, it is cataclysmic. That is why comets, along with asteroids, are closely watched by scientists and astronomers.

There are celestial displays that are much closer to earth, and much less portentous than comets, though they are still on the edge of space. Among them are the northern lights, which are probably the most impressive of all nocturnal spectacles. I have been privileged to see all sorts of auroral displays in northern Canada, from slowly undulating curtains of green and pink to flashing electric light shows that filled the entire sky. The aurora dances to the music of charged particles from the sun streaming into the earth's poles around the curve of earth's magnetosphere, and viewed from space, the aurora sits on top of the earth like a phosphorescent crown of light. The power of the aurora's charged particles is extraordinary and it has been estimated that a single display can generate a million megawatts of power, one hundred times greater than the few thousand megawatts of power generated by Niagara Falls.

According to the Vikings, the aurora borealis was the light of young women's souls and once, during the Middle Ages, a spectacular display of aurora borealis lasted for several nights and terrified the population of northern Europe. I can see why: an active display of northern lights turns the sky into a light show as quick and ever-changing as any light show at the height of the psychedelic era. The only other high-altitude nocturnal displays that come close are the noctilucent clouds, which are quite rare. I have never seen them. They are the highest clouds known, occurring at about sixty miles above the earth, usually just after sunset in northern latitudes. Observers claim they look like fairy clouds, appearing as bluish-white, irregular, sheetlike clouds with gold edges when they are near the horizon. Noctilucent clouds are probably formed of meteor dust and they move at terrific speeds, sometimes as fast as 394 miles per hour.

We have to take the relative distances of all these phenomena on faith. Atmospheric clues can extend our sense of depth a few miles beyond the five-hundred-foot (two-hundred-meter) limit of our binocular vision, but when it comes to distances of hundreds, thousands, and millions of miles, space looks two-dimensional, and we have to use geometry to determine the relative distance of the stars and planets. Yet there are night skies where I could swear I saw immense distances, nights so deep that the earth seemed to teeter on the infinite regress of outer space. On these dizzying cobalt blue nights, where there is no "up" or "down," it feels as if your body dangles into space with only your feet stuck to the earth. On such nights it seems a puff of air could launch you into the stars.

11

INSOMNIA—3 A.M.

"To Sleep"

A flock of sheep that leisurely pass by,
One after one; the sound of rain and bees
Murmuring; the fall of rivers, winds and seas,
Smooth fields, white sheets of water, and pure sky;
By turns have all been thought of; yet I lie,
Sleepless, and soon the small birds' melodies
Must hear, first uttered from my orchard trees;
And the first cuckoo's melancholy cry.
Even thus last night, and two nights more, I lay,
And could not win thee, sleep! by any stealth:
So do not let me wear tonight away:
Without thee what is all the morning's wealth?
Come, blessed barrier betwixt day and day,
Dear mother of fresh thoughts and joyous health!

WILLIAM WORDSWORTH

THREE A.M. IS THE heart of night. It is night's night, the innermost ring of darkness. Midnight might be the mathematical center of our ideal night, but 3 A.M. is later and somehow darker, it feels deeper into the night. Several hours of darkness remain ahead and yet the night is already nine hours old. "3 A.M. Eternal" is the title of a popular song by the nineties techno-rap group the KLF. According to the KLF, 3 A.M. is the hour of evening that best represents the eternal nature of time, and it is a rite of passage for those who intend to dance all night long. At 3 A.M. everyone who is not working or dancing is asleep, and if they aren't asleep, they are lying awake in their beds wishing that they were.

In Gabriel García Márquez's *One Hundred Years of Solitude* there is a story within the novel about how all the inhabitants of the village of Maconodo become insomniacs, or rather, they lose the need to sleep. As a result they continue their regular affairs night and day. Everything seems normal at first, but gradually they all begin to lose their memories. García Márquez was probably aware of the intimate relationship between sleep and memory and used this knowledge to lend an element of veracity to what would otherwise seem to be an arbitrary narrative. Sleep researchers have discovered an incontrovertible link between memory dysfunction and lack of sleep, particularly when it comes to the ability to recall names and proper nouns.

> *Tired Nature's sweet restorer, balmy sleep!*
> *He, like the world, his ready visit pays*
> *Where fortune smiles; the wretched he forsakes*
> EDWARD YOUNG

Lying in a dark bedroom waiting for sleep to come can be like spending the night in an insidious torture chamber. Time slows down to a crawl and every sound, from the ticking of a clock to the hum of a refrigerator down the hall, seems amplified and portentous. But the dominatrix of insomnia is the bedside clock—the boredom and frustration of the insomniac is implacably measured out by this chronometer, apportioning the misery of a sleepless night second by second, minute by minute. There is something even more disturbing about the nature of time in a sleepless bedroom because the agony of insomnia arises from a paradox: namely, even though each minute feels interminable, and as a result it would seem the night might last forever, the insomniac's frustration and anxiety increases with every glance at the clock because the insomniac is racing against

time. Every hour not spent sleeping deepens the anticipated exhaustion of the next day. (Although, if the insomniac is not alone, he or
she can avail him- or herself of the lovers' axiom—that two hours of
lovemaking equals four hours of sleep.) If it's already 3 A.M. and you
are still awake, then you have less than four hours of available slumber
time left, and the tension mounts. You can't stop thinking about
being recently passed over for a promotion at work. What if that
blood test turns out to be positive? Who is that person who keeps
calling, blocking their number and leaving no message? Welcome
to insomnia.

Insomnia

Only when one cannot sleep does one know how long the night is.
CHINESE PROVERB

To sleep at night is a natural thing, something most of us can count
on every evening; in fact, many of us have trouble keeping awake
when we're tired. We fall asleep at films and board meetings, we
catnap and siesta. But for some 10 percent of the population, sleep is
maddeningly elusive. In addition, our modern lifestyle has institutionalized a type of artificial insomnia created by a world that is
increasingly "24–7." Electric lighting has created an artificial day in
the middle of the night. It's no coincidence that Thomas Edison, the
creator of the incandescent light that has done so much to eradicate
sleep, was himself an unrepentant, nay willful, insomniac. He once
wrote, "Most people overeat 100 percent and oversleep 100 percent.
That extra 100 percent makes them unhealthy and inefficient. The
person who sleeps eight or ten hours a night is never fully asleep and
never fully awake."

I don't know what oversleeping 100 percent means exactly—100 percent more than what? Should we not sleep at all? But he does make a point, one that is quite compelling for those among us who wish to be more productive—because who hasn't, at some time in their lives, wished they could stay awake in order to complete some important task? The pace of modern life conspires to eat up more and more of our time, and we juggle increasing responsibilities. It's not surprising that insomniacs don't really stand out in our twenty-four-hour society—there are so many diversions and nighttime activities that insomnia has become institutionalized. But there's a price. Workaholic-related sleep deprivation and a plethora of sleep disorders are burgeoning into an epidemic of insomnia. According to the director of the Stanford University Sleep Disorders Research Center, Dr. William Dement, "We are a sleep-sick society." In his best-selling book *The Promise of Sleep*, he warned that "Sleep quality is one of the key components in determining how well or ill people feel in general." But as the standards of material success keep climbing, the workaholic lifestyle is burgeoning. A recent Japanese study found that between 1970 and 1990 the average length of a night's sleep had dropped by thirty minutes.

It is not known exactly how many individuals suffer from insomnia in industrialized nations, but surveys report that 50 percent of adults have had brief periods of insomnia, while 10 percent have experienced acute (temporary) insomnia and 5 percent suffer from varying degrees of chronic insomnia. Of the borderline cases that fall between acute and chronic insomnia, 24 percent are women and 18 percent are men. Recent literature suggests that this higher figure for women is partially due to the hormonal fluctuations of their monthly cycle. As people age, their susceptibility to insomnia increases: 50 percent of the elderly worldwide suffer from some form of insomnia. Insomnia also seems to be endemic among driven and accomplished people. The list of famous insomniacs is extensive, though they seem to be grouped into three basic occupations: politicians, artists, and inventers. In the political category are Margaret Thatcher (who once said, "Sleep is for wimps"), Winston Churchill, Napoleon Bonaparte, and Catherine the Great. Cary Grant and Marilyn Monroe certainly qualify as thespian artists, and the painter Vincent van Gogh was an inveterate insomniac. Entertainers also have a high predisposition to insomnia. In Christopher Anderson's biography of Madonna, *Madonna Unauthorized* (published in 1991), Madonna briefly sketches out the constellation of symptoms peculiar to ambition and accomplishment: "I'm anal retentive. I'm a workaholic. I have insomnia. And I'm a control freak. That's why I'm not married. Who could stand me?" Although she did eventually marry, the association of insomnia and compulsive behavior is clear. Thomas Edison and Benjamin Franklin round out the inventors list, but it seems that the lion's share of insomnia has been meted out to writers. The roll is long and illustrious: the litany of literary night owls includes luminaries like Marcel Proust, Emily Brontë, Colette, Evelyn Waugh, Virginia Woolf, Mark Twain, William Wordsworth, Alexandre Dumas, and Charles Dickens.

Perhaps the reason for the disproportionate number of literary insomniacs is because writers have been more voluble about describing their sleepless conditions. We can thank Emily Brontë for the adage "A ruffled mind makes a restless pillow." Perhaps she also transferred her anxiety about insomnia to her characters; after all, Heathcliff was something of a Gothic insomniac, perpetually pacing the nocturnal landscape of Wuthering Heights. Colette, with typical perversity, embraced sleeplessness. She wrote, "In its early stages, insomnia is almost an oasis in which those who have to think or suffer darkly take refuge." Mark Twain, although he suffered badly from insomnia, offered his own irreverent advice for falling asleep: "Try lying on the end of the bed, then you might drop off."

There is no doubt that Napoleon went over military strategies as he lay sleeplessly in bed, just as it is likely that Margaret Thatcher obsessed about election campaigns. But many famous insomniacs linked their sleeplessness not to their hyperactive minds, but to their surroundings. Winston Churchill, Benjamin Franklin, and Charles Dickens all connected their insomnia to the beds they slept in. Churchill had two beds—if he couldn't sleep in one, he hopped into the other. Franklin thought he couldn't sleep if his bed was too warm, so he'd get out of bed, roll down his sheets, and open a window until the bed was cool enough for him to resume sleeping. Joseph Stalin had an identical bedroom constructed for him wherever he went in Russia, right down to the wallpaper. For some people it seems that worry and sleep just don't seem to mix; for others, sleep remains elusive unless they have ideal conditions. But there are other reasons for insomnia, and many different kinds.

Types of Insomnia

Insomnia is a complex disorder with many causes. It is clinically divided into two broad categories by sleep disorder researchers: sleep onset insomnia and sleep maintenance insomnia. Sleep onset insomnia, which all of us have probably experienced, is characterized by difficulty falling asleep after lying down for the night. By contrast, sleep maintenance insomnia, more common in the elderly, is the inability to fall asleep again after waking up during the night. These two main types of insomnia are divided into two further categories, acute and chronic. Acute insomnia lasts less than a few weeks or months at most and usually ends without requiring treatment. Chronic insomnia lasts more than three months and almost always requires medical treatment.

One of the most common forms of sleeplessness is psychophysiological insomnia. People with this problem have difficulty sleeping because their general state of alertness or arousal is higher due to stress or illness. They have internalized their tensions, and as long as the stress factors are not resolved, they have difficulty sleeping. If this insomnia lasts long enough, then habitual nightly activities connected with sleep such as brushing teeth and turning off lights become associated with the source of stress and the insomniac enters a vicious cycle. People with this type of sleeplessness, which can last for years, often find they sleep better in unfamiliar environments, such as hotel rooms and couches. Another type of insomnia is sleep-state misperception, where sufferers believe they have slept much less than they actually have. Sleep-state misperception is a sort of psychosomatic insomnia, though people suffering from sleep-state misperception generally do have shorter, more restless sleeps as a result of their disorder. Sleep apnea, which was only recently discovered by a sleep laboratory in Paris in 1965, is another variety of insomnia, only more insidious because, unlike sleep-state misperception, the sleeper is often awakened many times throughout the night by obstruction of the airway, though he or she won't remember these many small awakenings upon getting up, feeling tired, in the morning. Five percent of the population have such severe sleep apnea that they have trouble staying awake during the day. People who suffer from sleep apnea are involved in 50 percent more car accidents than the general population and severe sleep apnea can cause even more insidious physical harm—it has been firmly linked to cardiac disease.

Some people get altitude insomnia (unrelated to jet lag) when they ascend to altitudes higher than four thousand meters (two miles) above sea level. Adjustment insomnia comes from trauma or temporary stress and new situations. Children sometimes suffer from limit-setting insomnia if their parents don't set regular bedtimes. Children can also develop sleep onset association disorder if they rely too heavily on special conditions for sleep, such as a pacifier or a television left on in another room. Infants occasionally suffer from food allergy insomnia, which generally occurs between the ages of two and four. Medications and their withdrawal can also cause insomnia, and anyone who has flown any distance has probably experienced jet lag insomnia, which comes from the difficulty of resetting our biological clocks in different time zones. This is the same problem that shift workers have when they alternate shifts, a sort of stationary jet lag. (We all suffer this to a certain extent in those areas that have daylight saving time. When the clocks are shifted an hour back in the fall and an hour forward in the spring, there is a slight peak in the number of motor vehicle accidents because so many drivers are technically, if mildly, "jet-lagged.") Delayed sleep-phase syndrome is a complex, organically based type of insomnia arising from a discrepancy between the sleeper's circadian rhythm and conventional sleep patterns. They find it difficult to awaken early in the morning and have trouble falling asleep before 2 A.M. Its opposite form of insomnia is advanced sleep-phase syndrome.

The most extreme form of insomnia is a very rare syndrome called fatal familial insomnia, or FFI. The most widely studied victims of fatal familial insomnia are members of an aristocratic Italian family who live near Venice. The syndrome, which seems to be transmitted genetically from one generation to the next, has plagued the Roiter family since 1836, when their ancestor Giacomo became the first member of the clan to die from what was, at the time, diagnosed as a mysterious illness characterized by total insomnia. Giacomo's descendants were very successful and wealthy, but many were afflicted with the same severe insomnia that caused premature death. Twenty-five percent of the family carry the disease-causing gene.

Typically FFI strikes its victims suddenly, usually when they are in their early fifties. One night they are unable to sleep. They try to nap the next day but just can't seem to doze off. The second night is also sleepless and by the third day their pupils become pinpoints, their blood pressure elevates, and their pulse begins to speed up. Sleepless weeks drag into months and yet their brains remain active, awake.

Eventually the victims lose the ability to speak, even to walk, though their minds remain conscious until, toward the end, they pass into a comatose state of terminal exhaustion. In terms of human suffering, FFI must be one of the most excruciating ways to die.

The exact cause of the disease was a mystery until 1997, when a researcher from the University of California, Stanley B. Prusiner, discovered a link between prions and FFI. Prions are rogue proteins that cause degenerative brain disease, of which the most famous is mad-cow disease. Creutzfeldt-Jakob disease, the human version of mad-cow disease, seemed to be related to FFI. Prusiner was able to prove that even though prions cannot be passed on genetically, successive members of FFI families have a specific helper protein that is inherited. It is the helper protein that triggers the prions to begin their pathological attack on the victim's brain.

Some researchers think that, despite the rarity of the disorder, curing FFI might also provide a cure for Creutzfeldt-Jakob disease and perhaps even Alzheimer's disease. William Dement of Stanford's sleep disorder clinic thinks FFI research should be massively funded, not just because of its similarity to other neurodegenerative diseases, but also because it might hold the key to curing insomnia, or aid in the creation of a new generation of safer and more efficient sleeping pills.

Wrestling with Angels: Treatments for Insomnia

Nothing cures insomnia like the realization that it's time to get up.
ANONYMOUS

Because there are so many forms and causes of insomnia, no two cases can be treated in the same way. Depression can cause insomnia and Prozac can be used to cure both the depression and the insomnia, for example, but in the end there is no magic pill for all types of insomnia. Tranquilizers end up being habit-forming and sleep aids can lose their effectiveness. A new generation of low-toxicity sleep inducers (the non-benzodiazepene hypnotics) shows some promise, though experts agree that any solution to insomnia that doesn't include medication is better than one that does. That is why, if you visit a sleep clinician with a diagnosis of chronic insomnia, he or she will first look at what they call your "sleep hygiene."

Good sleep hygiene for insomniacs includes standardizing their wake-up time, even on weekends. Eliminating nicotine, caffeine, alcohol, and other stimulants from the diet seems to help also. Caffeine is a well-known sleep disturber but alcohol can be equally debilitating. A nightcap may relax someone for sleep, but as alcohol is metabolized by the body, it releases carbohydrates and sugars that can result in night awakening hours later. Cessation of napping is often recommended, too, because even though catnaps seem particularly restful for the sleep-starved insomniac, they interfere with regular sleep rhythms. Exercise turns out to have a soporific effect, as does a hot bath before bed, and those who suffer from sleep-phase insomnia can benefit from taking melatonin four hours before bedtime. Insomniacs should limit their bed activity to sleeping only; they should avoid making phone calls, watching television, or listening to the radio in their beds. It is also best not to eat or drink just before going to bed because digesting food can cause wakefulness and a full bladder is a well-known interrupter of sleep. Maintaining a quiet sleep environment with a comfortable temperature and low-level lighting is also very important and, in some cases, calming techniques such as meditation or relaxation-sound tapes can help as well.

Some of the more aggressive techniques of combating insomnia include sleep restriction therapy, where the insomniac is encouraged get out of bed if still awake after twenty minutes of lying down. They are instructed to go back to bed only when they are truly tired. Of course, if natural techniques don't work then there are always medications. For temporary insomnia "sleep aids" can be helpful, though most of the over-the-counter sleep aids contain antihistamines, which can cause dry mouth, morning sickness, blurred vision, and decreased memory retention the next day. In terms of prescription medications the most effective sedatives are the sleeping pills, the benzodiazepenes. They are fairly low risk, though they can cause dependency, especially if taken for recreational use along with alcohol. In the film *Valley of the Dolls* (1967) Sharon Tate played a woman dependent on Valium, an early sleeping pill. A second generation of sleeping pills, including non-benzodiazepenes like zolpidem and zaleplon, have fewer side effects and are being increasingly prescribed for insomnia.

Famous Insomniacs

William Wordsworth was a voluble insomniac, and he often opined in verse about his condition. His lyrical complaints had the tone of bitterness and victimization that characterizes many sufferers of chronic insomnia. In one of his three poems on insomnia, all of which were titled "To Sleep," he rails against the elusiveness of slumber and the torment of sleeplessness: "This tiresome night . . . I am cross and peevish as a child." In another he addresses sleep itself: "Shall I alone, / I surely not a man ungently made, / Call thee worst tyrant by which flesh is crost? / Perverse, self-willed to own and disown, / Mere slave of them who never for thee prayed, / Still last to come where thou art wanted most!"

In Charles Dickens's *The Uncommercial Traveller* the author describes how his insomnia provided an incentive to go for walks at night around his favorite city. Dickens was the ultimate peripatetic writer and he took long regular walks both during the day and at night. Researchers have found evidence in his diaries that he suffered from acute sleep maintenance insomnia. He also must have had difficulty falling asleep because he had very exact requirements for a sound night's rest. He could only sleep in a bed that was aligned on a north-south axis, with the head of the bed pointing due north, and then only in the exact center of the bed, a position he checked on by extending his arms out sideways and measuring how centered he was on the mattress. He often awoke at night and went for walks. In one essay in *The Uncommercial Traveller*, he describes how he got up at 2 AM. and ended up "walking thirty miles into the country to breakfast. The road was so lonely in the night that I fell asleep to the monotonous sound of my own feet, doing their regular four miles an hour." In another essay in the same collection titled "Night Walks" he wrote about wandering the twilit streets of London. "Some years ago, a temporary inability to sleep, referable to a distressing impression, caused me to walk about the streets all night, for a series of several nights." Dickens put his insomnia to good use—he used his nocturnal excursions as social research forays into the conditions of the homeless in London, a cause he eventually championed in his novels. To him it seemed a great calamity that there were so many impoverished, homeless people taking what shelter they could: "Now and then in the night—but rarely—Houselessness would become aware of a furtive head peering out of a doorway a few yards before him, and, coming up with the head, would find a man standing bolt upright within the doorway's shadow, and evidently intent upon no particular service to society." Dickens believed nocturnal London was hiding a shameful secret and he wrote that "The wild moon and clouds were as restless as an evil conscience in a tumbled bed, and the very shadow of the immensity of London seemed to lie oppressively upon the river."

Marcel Proust, a chronic insomniac and nighthawk, usually stayed awake and wrote all night and slept, though fitfully, during the day. He kept his bedroom windows closed day and night, with the drapes pulled, and the walls of his bedroom were lined with cork for soundproofing. Proust availed himself of every conceivable potion and pharmaceutical that would cure his insomnia and during his worst episodes he tried opium, the sedative veronal, and even morphine to help him sleep. It's astonishing that he was able to achieve such lucid and transcendent narratives, with their subtle irony and humor, considering that, along with his insomnia, he suffered from asthma, hypersensitive skin, chronic dizziness, and an irrational fear of mice. Another sleepless author, Evelyn Waugh, used bromides to fall asleep, though his sleep was sometimes replaced by disturbing hallucinations as a side effect of his drug-induced stupor. Vincent van Gogh soaked his pillow and mattress with camphor, a type of aromatic oil, to help clear his mind of the deranged thoughts that plagued him as he tried to doze.

More recently the tragic story of Marilyn Monroe underscored the dark side of sleeplessness. The year after Marilyn Monroe finished shooting *The Misfits* in Reno, Nevada, her severe form of sleep onset insomnia was further exacerbated by psychophysiological insomnia. She had been involved in several, simultaneous relationships with famous men throughout the last decade of her life and the stress of these multiple relationships had brought about the breakdown that interrupted the filming of *The Misfits*. She had sought treatment for her depression and by the spring of 1962 she was having daily sessions with a psychiatrist. But her insomnia continued to worsen. "Nobody's really ever been able to tell me why I sleep so badly, but I know once I begin thinking, it's good-bye sleep," she once admitted to a reporter. "I used to think exercise helped— being in the country, fresh air, being with a man, sharing—but sometimes I can't sleep whatever I'm doing, unless I take some pills. And then it's only a drugged sleep. It's not the same as really sleeping." To alleviate her insomnia she took several prescription barbiturates and sedatives, some- times in combination. Her pharmaceutical tool kit included Sulfathalli- dine, Librium, and the phenobarbital Nembutal. In the last year of her life it was thought that she was taking up to twenty Nembutals a day.

Although some acquaintances thought her death was suicide, the consensus was that it was brought about by an accidental overdose in combination with alcohol. In a sense, insomnia killed Marilyn Monroe.

Insomnia is debilitating. Archibald Rosebery, who was prime minister of England from 1894 to 1895, was forced to resign prematurely during a period of severe insomnia. He wrote: "I cannot forget 1895. To lie, night after night, staring wide awake, hopeless of sleep, tormented in nerves . . . is an experience which no sane man with a conscience would repeat." After a bad sleep most of us lose our ability to concentrate and to remember names. After one or two nights without sleep we become stumbling fools, but without sleep entirely, we die, as do those suffering from FFI. Still, after all the research and discoveries, the basic reason for sleep continues to elude science. Does it facilitate the consolidation of memory, or does it help us forget useless information? Does it assist the immune system? Is it essential to bodily regeneration? Allen Rechtschaffen, a prominent American sleep researcher, wrote that "If sleep does not serve an absolutely vital function, it is the greatest mistake that evolution ever made." That sleep might be a mistake Edison would wholeheartedly agree, but Wordsworth and most of the rest of us would not.

The Blues: The Psychology of Night and the Curfew of Darkness

Insomnia aside, our perception of the passage of time at night is not the same as it is during the day. In daylight the position of the sun and its movement across the sky marks the progress of the hours, as do the shadows that it casts as they slide across floors and streets. Even on cloudy days, daytime schedules keep us busy and serve to indicate the motion of time. But at night, time behaves differently; it slows down, seems to stop sometimes, or jumps ahead inconsistently. Time at night is much more subjective. True, the transit of the moon across the sky does indicate the progress of time like the sun does, but the moon spends almost half of its monthly cycle out of view. As well, we pass most of our time at night inside, and indoors the steadiness of electric lighting gives our nocturnal interiors a timeless, constant illumination.

Sometimes the passage of nocturnal time is only apparent if we leave a room for a few hours and return. We come back to see that the leaves on a stalk of celery left on a table have wilted, or that the once hot bathwater is now cold or a puddle of coffee at the bottom of its cup has dried out. If the day is marked by constant change, if lunch is eaten and the business of the day gets done, then night is marked by an eerie perpetuality, particularly within the circle of artificial light. Indoor nights are slow-motion theaters in which time is an invisible actor.

This slowing down of time, or at least this alteration of time, heightens our sensitivity to the ebb and flow of our emotions. If we are lonely then night is when we feel most alone. And if we are frightened or apprehensive, night heightens our anxiety. But it is particularly when we are sad that night can turn our melancholy into deep blues and our depression into despair. Night restricts our liberty, not only because we are not as free to move as during the day, but also because we cannot see the larger space in which our liberty exists. Night imposes a sort of universal, low-grade curfew. The isolation that evening brings, along with the impairment of our vision by darkness, are felt as real, tangible losses.

Despite artificially lit nocturnal environments, our freedom at night is still limited to where electric light shines, or where we can carry our own light. This means that our loss of vision, even if temporary and reversible with the flick of a switch, is still, in an unconscious way, a handicap that underlies our sense of unease at night. A universal blindness descends upon us that has nothing to do with whether or not we're sighted. Perhaps that is why, even in our overly lit world, we still go through all the stages of denial, withdrawal, and depression, be they ever so subtle, that the loss of a sense entails. During the longer nights of winter those who suffer from seasonal affective disorder (SAD) have an even rougher time. The extent and severity of these symptoms varies with each individual, but an emotional reaction to our loss of sight is inevitable. Some people withdraw into themselves, losing motivation; others deny that anything has changed. It really depends on how much company we have, or at least the feeling of support. People in a family setting, or in a relationship, or even those with a pet can deal with night blues much more successfully than individuals without such support. Television fills the gap left by the absence of friends or family, and many single people spend their evenings in front of the TV screen. But television and the Internet are virtual, empty communities, and nocturnal depression can only be temporarily staunched by these electronic diversions. It is a widely known fact that the majority of suicides take place at night.

Night can be a time of memory, regret, and reminiscence. There are certain nights when we are alone in the darkness, perhaps lying down to sleep or sitting quietly by a window, that it seems the faces of loves lost, of departed friends, or regrets and remorse flood into our restless minds and overwhelm us. Remorse, particularly, seems grittier in the darkness, as T. S. Eliot wrote in his poem "La Figlia Che Piange" about the final parting of two lovers, remembered by the narrator of the poem as he sits alone in his room reminiscing at night: "Sometimes these cogitations still amaze / The troubled midnight . . ." Perhaps it is not just the lessening of vision but also the silence at night that evokes remembered faces, pressed like phantoms against the dark window of our memory.

Eliot was a master of this ambivalent, sometimes disturbing time of the soul. In "Rhapsody on a Windy Night" he wrote how "Memory throws up high and dry / A crowd of twisted things," and goes on to describe a tormented night of self-examination where, as he walks restlessly through the streets, the craters and seas of the moon look like "washed-out smallpox" and how each streetlight "Beats like a fatalistic drum"; finally, in the end he returns alone to his home at 4 A.M. The nighttime blues have created a whole genre of music, as well as some very sad country-and-western songs. In Hank Williams's "I'm So Lonesome I Could Cry," he describes losing the will to live at night, yet the song is peopled with wonderful, lyrical images of falling stars against a purple night sky and sung with such visionary, almost narcissistic sadness that it takes the night blues into another realm entirely, into a cosmic, lonely, self-pitying nostalgia that transcends depression and loss. If night is when we are susceptible to melancholy, it is at night we have to surmount our despair, to realize the darkness is not a barren exile, a period of house arrest, but filled with opportunities for absolution.

12

ENDLESS NIGHTS—4 A.M.

THIS IS THE last hour of deep night. Even the field crickets are slowing down—their chirps becoming more sporadic as the temperature dips below 50 degrees Fahrenheit. It is a clear evening and beads of dew in the grass are starting to amalgamate like small, shimmering amoebas. Sleepers are entering their second-to-last sleep cycles and their body temperatures have almost reached the lowest point of the evening. In the dreamless sleep at the bottom of their Delta rhythm, where brain-wave frequency slows to three cycles a second, sleepers are prone to heart arrhythmias and irregular breathing—and a few of them are being stalked by death. As Hesiod wrote in his *Theogony* in the eighth century B.C., "There dwell the children of the dark Night, the dread gods Sleep and Death." Cortisal, which is just beginning to be secreted by their adrenal glands, adds another layer of stress to their unconscious bodies.

In Las Vegas, above the corona of neon forest, the desert night seems to stretch into space, and from the rooftop of the Sands it appears that the stars are closer. The Stardust casino hosts a few all-night gamblers who are still riding high: a woman with bleached-blond hair plays a one-armed-bandit while a high-roller sits at the blackjack table, a pile of chips nestled in the crook of his arm and a highball in his right hand. You wouldn't know it to look at him, but he is a reverse vampire. Earlier in the day he ran out of credit and sold his blood to a downtown blood clinic to work up a little stake for the crap tables. His sanguineous gamble paid off and now he's hooked on an all-night run of luck.

Las Vegas is the ultimate desert flower—blooming suddenly, brightly colored, and night-perfumed. Less than a hundred years ago Las Vegas was a handful of buildings and a railway station. Now its population has swollen to 1.25 million residents who enjoy a booming economy driven by entertainment, gambling, and tourism. Perhaps Las Vegas poses a glimpse of America's future, as more and more economists predict that an entertainment economy will become the dominant engine of wealth in the Western world. Glitter Gulch is an oasis of luminous media in a desert landscape, and, as if to underscore the metaphor, water fountains are everywhere, like a Persian desert dream. In Las Vegas the glamour and excitement of night is perpetuated indoors. French cultural critic Jean Baudrillard described Las Vegas in "Astral America" in 1986: "The skylines lit up at dead of night, the air-conditioning systems cooling empty hotels in the desert and artificial light in the middle of the day all have something both demented and admirable about them. The mindless luxury of a rich civilization, and yet of a civilization perhaps as scared to see the lights go out as was the hunter in his primitive night."

Baudrillard may have been right in general terms about America's nyctophobia, but I'm not sure he really understands Las Vegas's perpetual, artificial day. It may seem strange but, in some paradoxical sense, the "day" of Las Vegas is really a night in lucent disguise—for most gambling takes place at night, and in Las Vegas that "night," even though it is bright with synthetic light, seems endless. That is why Las Vegas casinos have no windows and no clocks on their walls. In Las Vegas even night itself has been "themed." Darkness is just another aspect of a completely simulated environment. The arcade attached to Caesars Palace, called Caesars Forum, is a life-size reconstruction of an old European arcade, its winding streets filled with shops and tourists. Above the streets specially illuminated domed archways re-create the sky, gradually changing from day to evening every twenty minutes. The rhythms of Las Vegas are definitely not circadian. They are the arbitrary rhythms of a possible human future where endless artificial night creates a new environment that could, if necessary, exist in a cavern beneath the earth, like the underground, Midwestern town featured in the 1975 movie *A Boy and His Dog*, or perhaps on a large spaceship traveling between the stars in some science-fiction odyssey. Though Las Vegas perpetuates its desert night with simulated dreams, there are other places in the world where darkness lingers longer without any assistance from us.

The Polar Night

Although not technically endless, polar nights are the longest on earth. Because our planet's axis is tilted about 23.5 degrees off its orbital plane, the number of hours of sunlight each day is diminished or increased according to the position of the earth within its yearly, six-hundred-million-mile orbit around the sun. Our seasons of winter, spring, summer, and fall are a result of the earth's tilt, as are the long nights of the polar regions. When the Northern Hemisphere is tipped away from the sun, in winter, night reigns. The Arctic Circle, at latitude 66°33' N, marks the exact point where the tilt of the earth is so extreme that at least on one day of the winter the sun never rises. For each degree farther north of the Arctic Circle the length of the polar night increases proportionately until, at the North Pole, it reaches its maximum length.

In Grise Fiord, a port on the south coast of Ellesmere Island well within the Canadian Arctic Circle, the sun disappears on October 31 and is not seen again until February 11, more than a hundred days later. In Alert, the northernmost settlement in the Canadian Arctic on the northern tip of Ellesmere Island, the polar night lasts more than 130 days. But just a little more than five hundred miles north of Alert, at the North Pole, the polar night lasts 179 days. In fact, the North and South Poles experience only one day and one night a year, each 179 days long. Although today the lands within the Arctic Circle are frozen there was a time, long ago, when the Arctic was covered with great forests.

The Endless Summer Night of the Arctic Eocene

On the west side of Ellesmere Island the straits of Eureka Sound separate it from its sister island to the west, Axel Heiberg. Both islands are large, with low mountainous topography supporting barren Arctic deserts. The tallest trees are a type of dwarf willow that creeps along the ground rising no more than two or three inches into the air. It wasn't always this barren. In the claylike sedimentary deposits that straddle Eureka Sound, several bands of coal-dark strata trace the contours of the foothills. These strata, now called the Eureka Sound Formation, were first discovered in 1956 by Neil McMillan, who worked for the Geological Survey of Canada. He dated the Eureka Sound Formation at forty million to fifty million years old, which put it in the Eocene period, or the age of early mammals, a few million years after the dinosaurs had died out. McMillan also noticed that these coal-like layers were rich with fossils. News of his discovery eventually made its way through the geological community until, in 1972, it reached the ears of Mary Dawson, the curator of vertebrate fossils at the Carnegie Museum of Natural History in Pittsburgh and an expert in Eocene mammals. Dawson realized the importance of these Eocene deposits and decided to visit them to make a firsthand examination. She couldn't have predicted the series of startling discoveries that she was about to make.

Dawson went on several field trips to Ellesmere in the mid 1970s accompanied by Robert M. West, also from the Carnegie Museum of Natural History. Their first excursion, in 1973, was disappointing. They found no animal fossils. But on their second trip in 1975, they discovered not only the bones of horses, tapirs, and rhinoceroses, they also found the remains of tortoises and alligators—animals that could only exist in a frost-free environment where the average annual temperature never dipped below 60 degrees Fahrenheit—a subtropical climate. What was mind-boggling about their discovery was the fact that forty-five million years ago Ellesmere Island was only 5 degrees south of its present position, well within the Arctic Circle. This meant that the Eocene climate was included in a worldwide heat wave, where even the Arctic regions enjoyed a subtropical climate. Even more important, a very unique, subtropical ecosystem had evolved to flourish during a "winter" night that lasted for one hundred days, the same length of Arctic night that Gris Fiord experiences today on the south shore of Ellesmere Island. What they had unearthed were the remains of the longest subtropical nights in our planet's history.

Over the next two decades a host of geologists and paleontologists descended on the region. In 1985, Paul Tudge was piloting a helicopter for the Geological Survey of Canada over Axel Heiberg Island when, during a low-level pass, he saw a hillside strewn with logs and large stumps. He landed and realized that he'd found the best-preserved Eocene site yet, the remains of an ancient forest with hundreds of stumps and logs buried in a dark layer of leaf mulch, all of which turned out to be anaerobically preserved. None of it was fossilized even though it was more than forty million years old. News of Tudge's discovery brought teams from around the world to the site and gradually a much clearer picture of what it must have been like in the subtropical forests of the Arctic Eocene began to emerge.

The primeval forests of Axel Heiberg Island were dominated by 160-foot-tall dawn redwoods, a type of deciduous evergreen tree with a conical silhouette and feathery, soft leaves. Interspersed among them were towering swamp cypress trees. Because the land that these giants grew in was low and wet, the overall appearance of the Eocene forests on Axel Heiberg must have been similar to the bald cypress swamps of the American Deep South. Other trees in the fossil forest included ancestral hickories, birches, maples, and sycamore, all of which had huge leaves, often more than two feet across. Their leaves had to be large to take advantage of the Arctic summer. Some paleobotanists believe that deciduous trees originated in the Arctic Eocene, where they dropped their leaves each fall as a way of preserving precious energy during the long Arctic night. Browsing among the trees were herds of coryphodon, a sort of pygmy hippopotamus, as well as modern-looking tapirs. Other denizens included large varanid lizards the size of Komodo dragons, boa constrictors, giant land tortoises, four-foot-long salamanders, and a six-foot-tall flightless raptor called Diatryma, also referred to as the terror crane, a predatory bird that filled in the niche left vacant by the extinct velociraptors and tyrannasauroids.

But it was when the long darkness fell that things got interesting in the Eocene forests of Axel Heiberg. Because the winter was warm, the Eocene twilight was more like a three-month-long summer night. Some paleontologists believe that during the winter months these lowland cypress swamps were perpetually fog-shrouded, effectively insulating them against the low temperatures that would normally accompany such a long night. In this darkness it is thought that many of today's nocturnal mammals got their evolutionary start; if they were to be active during this period, those whose eyes gathered more light, whose hearing was more acute, flourished. For instance, from the copious number of lemur skulls disinterred by Dawson and other paleontologists on Axel Heiberg, it seems that the dark canopy of the Eocene forest rustled with the movements of many species of lemurs, including a flying lemur that could glide from tree to tree. Although no one has seriously proposed a theory that the first bats gained their evolutionary wings in these same forests, it is entirely possible, for the earliest bat fossils appear in strata of this age. Nevertheless it is certain that, at least by forty million years ago, bats flew through the perpetual gloom of this most exotic of primeval subtropical environments. As it is today, the aurora borealis was active during the Eocene, and on some evenings the glowing, shifting electric curtains of the borealis must have provided an eerie glow that dimly illuminated the protracted, subtropical twilight of the Arctic Eocene forest.

Cave Darkness

For Nature beats in perfect tune,
And rounds with rhyme her every rune,
Whether she work in land or sea,
Or hide underground her alchemy.

RALPH WALDO EMERSON

If the Arctic night is long, it is not technically endless. For truly interminable nights we have to go beneath the earth's surface, to the realm of the limestone caves. All limestone caverns are formed in the same way, in a geological process that takes millions of years. Rainwater, which is slightly acidic relative to limestone, seeps through crevices in the surface and begins to dissolve the rock underneath. This creates small channels that over the years inter-connect and after many centuries form underground rivers that create even larger caverns. Over the millennia, as water tables lower or the land rises, these once submerged passages become dry caves. Dissolved calcite dripping from the ceilings of the caverns solidifies into stalactites, like stone icicles, which lengthen and expand into immense rippled columns that sometimes form into irregular colonnades, like surreal underground cathedrals. All of these fantastic forms are hidden, like many of our planet's deeply buried wonders, in the geological night of the earth's subsurface.

Caves remain at a constant 55 degrees Fahrenheit, even in the coldest weather, for all the world's caves, including northern caves, are well below the winter frost line. (The only caves that are warmer are manmade ones. The deep diamond mines in South Africa have ambient temperatures of more than 80 degrees Fahrenheit because they are that much closer to the molten magma upon which the continents float.) Because the underground climate of caves is constant, they make a good refuge for animals. In the north bats take advantage of the warmth, and many animals find shelter in the entrance of caves. But there is another group of very specialized creatures that never leave the confines of the caverns.

Over the thousands and millions of years that it takes for caves to form, small creatures sometimes fall or swim into the caverns and are trapped there, and over the centuries those that survive and reproduce gradually evolve into strange, pale denizens of the endless night of deep caves. Fish, salamanders, crayfish, amphipods, and some insects have adapted to the darkness and the scarce food resources of underground caverns. After hundreds of generations most of these creatures have lost any pigmentation in their skin, and many have also lost their eyes. Though cave animals exist all over the world, some of the best-studied cave dwellers live in the big caverns of the southern United States.

The eyeless American cavefishes, for example, are small, four-inch-long codlike fish whose lives are restricted to the dark. They are completely blind and have a faintly pigmented, whitish-pink color. Their only source of food consists of tiny cave-dwelling invertebrates that the cavefish detect with specialized sensory papillae on their heads and bodies. A spelunker (cave explorer) from the National Speleological Society in Huntsville, Alabama, told me that young cavefish are almost completely transparent, and, like the rare quarter-inch cave shrimp they feed on, they can be detected only by the shadows they cast on the bottom of crystal-clear cave pools under a flashlight beam. There are also blind, colorless crayfish in these caves and blind, eyeless white salamanders. Of these, the most extreme salamander in the United States is the Texas blind salamander, a three-to-five-inch, very slender, semitranslucent milky amphibian with skinny, sticklike arms and a ruff of red gills that makes it look like a diminutive, naked haute-couture model sporting a red scarf.

Meramec Caverns

Although I've never seen any of these creatures myself, I *have* experienced the pitch-black geological night of a major cave. A few years ago I went on a walking tour of the Meramec Caverns in Missouri. I saw giant stalactites and stalagmites (pyramidal stone accretions that build up from the floor of caves) and in one cave, a row of large stalactites that had congealed together to form a continuous surface. They reminded me of the huge icicles that form at the base of Niagara Falls during the winter, only these rock versions were made of calcite shot through with magnificent mineralized colors—red, ochre, cream, white, and gray. At one point on our tour the guide suggested that we experience the total darkness of the underground, called "cave darkness," by turning off all our flashlights and remaining still for some time to allow our eyes to get used to the dark. An inky, claustrophobic blackness descended on us and after a minute or two it just seemed to get darker. There wasn't a single photon of light, we were immersed in a pitch darkness just a faulty flashlight away from panic. I realized that this was the kind of night only the real masters of the dark could negotiate. Passive night vision wouldn't work here, and most owls, except the barn owl, would be stymied by the total lack of light. Bats, with their echolocation system, could navigate easily, but as for me, without any light I'd be reduced to feeling my way along the rough, damp walls of the limestone passages.

The experience of cave darkness gave me renewed respect for spelunkers, particularly those of the past, who explored without modern equipment. I'm not completely free of claustrophobia, and squeezing through narrow passageways hundreds of feet below the ground gives me the creeps. A particular kind of bravery is required by spelunkers, and perhaps the bravest cave explorer I've heard of was the intrepid Norbert Casteret, a French speleologist (cave scientist) and archaeologist who risked his life to discover a cave containing a prehistoric carving in 1922. The carving he discovered was in a partially mapped cave system called the Montespan Caves in the Pyrenees. While exploring some of the lower passages, he found a new opening. He entered it and followed a passageway that got narrower and narrower as it descended until finally he had to stop because the bottom was flooded with water. Undaunted, he took his clothes off, put a waterproof match and a candle under his woolen toque, and dove into the cold water at the bottom of the passageway.

Submerged in the cold water in total darkness, and feeling his way forward through the narrow passageway (called a "siphon" by spelunkers, an underwater passage that sometimes connects one dry cave to another), he reached the point of no return; with his lungs close to bursting he knew that if he continued he would not have enough air to get back to his entry point. If there wasn't an opening close ahead he would drown alone, in total blackness, at the bottom of a deep cave. He decided to keep going. It was either an act of supreme courage or absolute madness, but there in the eternal night of the underwater passage, Norbert Casteret forged ahead, and his gamble paid off. The passage began to curve upward and he surfaced, gasping for breath, in a new cavern.

A year later, revisiting the same cave, he explored it more thoroughly and discovered a life-size clay sculpture of a headless cave bear (*Ursus speloeus*). Lying between its front paws was the skull of a real cave bear. It turned out that the figurine was part of a prehistoric cave-bear worship ritual. Giant cave bears have been extinct for twelve thousand years and his candle was the first light to break the endless subterranean night that had surrounded this figurine for more than twenty thousand years.

The Abyss

Night, pitch-black, lies upon the deep.
VIRGIL

Although the eternal night of deep caves supports a handful of translucent animals, there is another perpetual twilight that harbors an entire world of creatures adapted to endless nights. It exists in the lowest depths of the world's oceans—the realm of the abyss. These deep waters are home to some of the most alien-looking creatures on our planet, and they swim through a submerged domain that is three times larger than any continent. If our hypothetical visitors from space, the ones who watched the growth of city lights on the dark side of our planet, had approached earth from the Pacific Ocean side, they would have thought that earth was rather inappropriately named. From their viewpoint high over the middle of the Pacific Ocean, they would see no land at all, aside from a few islands and the edge of Australia—after all, only 29 percent of the earth rises above sea level, the other 71 percent is submerged.

We are tied to the sea, and its dark abyss, by blood. Life arose in the oceans and we still carry it within our bodies, which are 70 percent saline water. Blood, sweat, and tears are portions of the primordial ocean that cradled our ancestors. That physical link to the ocean means that, in a sense, what goes on within the oceans goes on within us. Perhaps that is why, as Carl Jung theorized, dreams of water must be dreams about our unconscious selves. The creatures within us that haunt the vast, twilight ocean depths of our own unconscious minds are cousins to the denizens of the abyss.

The deepest ocean trenches are farther below sea level than the highest mountains are above it. If you took Mount Everest, which is 29,000 feet high, and dropped it into the Marianas trench (just off the island of Guam in the south Pacific Ocean), not only would it be completely submerged, a mile of water would cover its summit. The bottom of the Marianas trench is 35,813 feet, or seven miles, deep. That's farther under the surface of the ocean than most commercial jet airliners fly above the earth. In this zone, where the water pressure is a bone-crushing sixteen thousand pounds per square inch, almost all living creatures are adorned with light-emitting organs. Because only 1 percent of the light that strikes the ocean's surface penetrates any deeper than three hundred feet, everything lower than that is immersed in eternal night, and the fish that patrol these cold waters between six hundred and five thousand feet produce their own light from light-emitting organs called photophores spotted along the lengths of their bodies like Christmas lights. Unlike bioluminescence, these organs collect and then reflect what scarce light they can gather. Similar to land-based nocturnal animals, deep-sea fish also have huge eyes to amplify the scarce light, and due to the great water pressure, the bodies of most of these abyssal pisceans are flattened from side to side, laterally.

There are many different species of fish that navigate the bathypelagic twilight two thousand feet beneath the surface of the ocean. In the abyssal gloom at the foot of the continental shelf, schools of lanternfish with vertical, angular bodies, huge eyes, and glowing photophores, swim in luminescent, flickering schools. They are hunted by some of the most sinister-looking fish on the planet, including Sloane's viperfish, a two-foot, blue-black horror with an eel-like body and a dispropor-tionately large mouth brimming with flexible, needle-thin, transparent teeth. Its teeth are so long that the viperfish can never completely close its mouth. In fact, disproportionately large, toothy mouths are the rule in the abyssal night of the deep. The snakelike black dragonfish has a mouth that also bristles with painfully sharp transparent teeth, set in jaws that can open to a 180-degree angle. But the slackjaw takes the prize for ravenous bites. Where the jaws of most other vertebrate animals have a hinge that holds them together, the slackjaw has no hinge—its jaws are completely independent of each other, enabling it to open its mouth so wide it can (and it does) eat prey larger than itself! (It also has an expanding, flexible stomach that can stretch like a balloon.) All of these predatory, deep-sea fish are essentially swimming mouths lined with diabolically sharp teeth. They make sharks look like nibbling novices, demure diners quietly sipping soup at the back table of a delicatessen. With photophores along the length of their bodies accentuating their snakelike swimming movements, these carnivores must seem like glowing skeleton-fish as they undulate toward their prey. Of the three mentioned above, the deepest swimmer is the viperfish, which is found worldwide to depths of five thousand feet, deeper even than the vampire squid (a recently discovered squid that has short tentacles and a large, swollen blue-white mantle with disturbingly human-looking eyes). Other gaping demons of the deep include anglerfish like the blackdevil, which bears an absurd resemblance to its Tasmanian cousin, with an oversized mouth and short, stumpy body. A relative, the bearded angler, has an elaborate, bioluminescent branching lure attached to its head that looks like glowing seaweed.

In some bathypelagic areas (two thousand to twelve thousand feet deep), close to where the continental plates are being ripped open and the midocean floor is parting, clusters of hydrothermal vents spew hot water and dissolved minerals into the ocean 2,300 feet below the surface. There, attracted to the heat and mineral deposits, even stranger creatures congregate: bloodred tube worms and white crabs living beside superheated water gushing out of calcified chimneys at 760 degrees Fahrenheit, though the ocean temperature just a few feet from the hot vents is only a few degrees above freezing. The depths of the ocean also seem to engender gigantism. The dark abyss is the realm of giant squids, some of which have reached more than sixty feet in length, and diving down into the gloom to hunt these huge mollusks are sperm whales who patrol the upper levels of the bathypelagic zone to chase their favorite prey. Sometimes, though, the giant squids are too big for even the sperm whales to handle, and the sucker scars left on some sperm whales attest to the ferocity of their deep-sea battles.

But the most lethal of the undersea giants are not creatures at all, they are the nuclear submarines that regularly patrol the ocean depths. During the cold war U.S. and Russian submarines engaged in dangerous games of brinkmanship in the deep twilight of the world's oceans, trying to outdo each other in submerged versions of stealth and blind-man's bluff. These machines were behemoths; the two cold war protagonists, which still prowl the depths today, were the U.S. Ohio-class strategic-missile-launching submarines, weighing in at 18,750 tons and at 548 feet in length, with a crew of 160 and 24 nuclear missiles (with enough punch to wipe out an entire continent), and the Russian Typhoon-class strategic-missile-launching sub-marines, weighing more at 26,500 tons and 10 feet longer than the giant American subs at 558 feet. The Russian subs could launch 20 nuclear missiles and had a crew of 175. Both of these subs ran so silently that sometimes they collided during their cold war games of stealth. Even after glasnost and the fall of the Berlin wall their war games continued. Unlike deep-sea fish, nuclear submarines have no "running lights" to warn their prey of their approach. They are menacing, cigar-shaped pods, dark on the outside and light-filled within. The Typhoons and Ohios are streamlined underwater build-ings with stairwells and bedrooms. In this sense submarines are like submerged capsules of light within the pitch night of the deep ocean, like architectural enigmas of the abyss. The reverse is true of architectures that capture night during the day on dry land.

Planetariums

After the great fire of Rome in A.D. 64, Nero built a palace called the Domus Aurea that was centered on a domed hall called the *coenatio*, where the emperor appeared during private rituals celebrating his deification. Above him, inside the domed ceiling of the *coenatio*, Nero commissioned two of Rome's foremost architects at the time, Severus and Celer, to build a series of nested, semispherical, rotating domes that simulated the movement of the planets in the night sky. As Roman historian Suetonius wrote, "The main chamber was round and revolved continually day and night, as does the world." Nero also had his engineers build an ingenious network of metal tubing that was incorporated into the mobile domes so that the effect of rain could be reproduced, which, on Nero's command, would fall from his artificial sky. Seneca, another Roman historian, who was Nero's preceptor, described how these mechanisms worked: "a technician invented a system by which saffron-colored water poured down from a great height and also succeeded in assembling the panels of this chamber's ceiling in such a way that it changed at will." Unfortunately for Nero, he was despised by the Flavian dynasty that supplanted him and his works were destroyed in a retributive policy of *damnatio memoriae*.

But the memory of Nero's great *coenatio* survived after all, and when Domitian built his private palace, the Domus Agustana, high on the Palatine hill overlooking the Circus Maximus in A.D. 80, he constructed his own *coenatio*, which he called Coenatio Jovis (Dining Hall of Jupiter). This was a huge hall, measuring 87 feet high by 99 feet wide, in which Domitian installed the same nested, rotating domes as Nero had in his *coenatio*. But Domitian took his simulated sky one step further and had glowing stars inset into the domes. The poet Statius described these as being "stars like those of the celestial vault." When Domitian appeared during ritual ceremonies in his *sanctum palatium*, he was surrounded by projections and magical effects that must have been awe-inspiring to the onlookers. Statius related in his *Silvae* that upon entering the sanctum he "found himself with Jupiter (Domitian himself) amongst the stars."

In a sense Domitian's *coenatio* was the world's first planetarium. It wasn't equaled for almost two thousand years, until, in 1924, in Jena, Germany, the Carl Zeiss optical works surpassed Domitian's planetarium by building the first star projector. The Zeiss star projector resembled a large, black dumbell with hundreds of small lenses inset into the surface of the "bells." But when it was turned on, each of those lenses accurately projected a star or planet in its relative position in the night sky onto whatever surface the projector was aimed toward, instantly creating the illusion of a night sky. As well, the whole projector could be rotated with electric motors to simulate the motion of the stars. The Zeiss star projector became the standard for planetariums around the world and by the 1960s the top-of-the-line model could project nine thousand stars and simulate the Milky Way. Anyone who has been to a modern planetarium can tell you how realistic the Zeiss night sky is. When the simulated sunset fades and the stars begin to shine on the planetarium's ceiling, you can almost feel the first cool night breezes playing upon your face.

The planetarium's illusion of night is so complete that when you leave a planetarium during the day the bright sunlight is shocking. It's the same feeling you get when you leave a movie theater during the afternoon. In his book *Varieties of Visual Experience*, Edmund Burke Feldman wrote about how the potent nocturnal ambience of cinemas not only evokes the experience of night, but also of sleep: "The darkened theatre; a soft, comfortable seat; controlled temperature and humidity; and thousands of shadowy images moving across the screen—these strongly suggest the dream world each of us occupies while sleeping. It is possible to think of cinema as the deliberate manufacture of dreams."

Like planetariums, movie theaters are also temples of the night, but the creatures that inhabit them are not the evolutionary consequences of continual darkness, they are products of our own imagination. Our theaters flicker with the images of our nightmares and our dreams, but our evocations of night are not only confined to movies and planetariums, they are also found in literature, art, and music.

13

THE ART OF DARKNESS—5 A.M.

We are the music-makers,
And we are the dreamers of dreams,
Wandering by lone sea-breakers,
And sitting by desolate streams;
World-losers and world-forsakers,
On whom the pale moon gleams

ARTHUR WILLIAM
EDGAR O'SHAUGHNESSY

IT IS 5 A.M., THE last hour of night. Soon the first scattered birdcalls will begin—sparrows and pigeons announcing the urban dawn while, in the country, crows and warblers will sound the first light. This is the legendary hour when vampires must find their lairs, when all-night parties break up, and exhausted dancers start to make their way home. In the city there is a brief juxtaposition of the first day workers and urban nighthawks, like an encounter of two alien cultures. A tall woman in a designer dress gets into the same taxi a stockbroker vacates, briefcase in hand. The place where he sat in the back seat is still warm and the woman slides over to a cooler section. She is going home to sleep, her condominium blinds drawn against the morning sun. The stockbroker will sip a second coffee while scanning the latest European market numbers on his computer.

Already the eastern sky is beginning to glow and the stars are fading. In their beds sleepers are more restless, while their core body temperatures have now cooled to their lowest point, 96 degrees Fahrenheit; their cortisol levels are starting to peak at their highest concentrations of the night. Soon alarms will begin to buzz, ring, beep, and trill and a new day will break. This is the hour of anticipation. As Walt Whitman wrote in his journal *Specimen Days* on July 23, 1878: "I have itemized the night—but dare I attempt the cloudless dawn? (What subtle tie is this between one's soul and the break of day? Alike, and yet no two nights or morning show ever exactly alike.)"

The last hour of darkness catches some by surprise; the artists who toil all night long immersed in their work are particularly vulnerable to a state of mind that Vladimir Nabokov was familiar with: "Here is what sometimes happened to me: after spending the first part of the night at my desk—that part when night trudges heavily uphill—I would emerge from the trance of my task at the exact moment when night had reached the summit and was teetering on that crest, ready to roll down into the haze of dawn."

What better time then, and what better way to savor the last minutes of twilight than to cast back over the cultural history of night, to survey the sounds and images that have formed our experience of darkness, and to see how artists and musicians and writers have changed our conception of night. One of the marks of great art is how permanently it affects our subsequent experience of what it portrays. Vincent van Gogh's paintings of stars, depicted as swirling pinwheels of light in the summer darkness over southern France, have conditioned us to see stars with all the visionary intensity he infused in them. Turner's paintings of sunsets, likewise, have had the same effect. Although the clearest images of night are in literature, paintings, and film, music has also shaped how we perceive darkness. In some ways music has given us the most accurate portrayal of night because it conjures nocturnal moods so marvelously. Music, like the night, is ineffable—it surrounds us yet cannot be grasped. There are many haunting classical compositions that have been inspired by twilight: Debussy's *Clair de Lune* is perhaps one of the most well known, certainly Mozart's extraordinary *Eine Kleine Nacht-*

musik is named for the beauty of night, and Gustav Holst's *The Planets* evokes the deep night sky within which Jupiter and Saturn spin in somber majesty like giant symphonic enigmas near the outer reaches of the solar system.

Similar to their literary contemporaries, the musicians of the late eighteenth and early nineteenth centuries were romantically inspired by night. It was around this period of time that the nocturne, a free-form musical arrangement invented by John Field in the early nineteenth century, became a favorite arrangement for many composers because of its languid, reflective character. Chopin composed nineteen nocturnes for the piano and Francis Poulenc and Gabriel Fauré also wrote several night pieces for piano. Schubert composed a collection of lovely nocturnes for two instruments including *Die Nacht ("Die Nacht ist dumpfig und finster")*. Of all the compositions that evoke night, Satie's *Trois Gymnopedies* strikes me as being the most nocturnal of all, perfectly embodying the enchantment and stillness of a summer evening. Tchaikovsky was a decidedly nocturnal composer whose symphonies straddled music and dance. His two most famous ballets, *Swan Lake* and *The Nutcracker*, both feature exquisite nocturnal dances set in late fall/early winter twilight—"Dance of the Swans" and "Dance of the Sugar Plum Fairy." *Giselle* is another nocturnal ballet, choreographed for the ghosts of girls who had died for love. In terms of opera, perhaps the most nocturnal is Bartök's *Bluebeard's Castle*, for the interior of Bluebeard's castle is shrouded in a timeless, magical night.

Popular songs have framed some of the most soulful and directly romantic visions of night, and my own vision of night was certainly shaped by them. When I was eleven I had a favorite 45-rpm record that I played incessantly. It was an instrumental version of "Harlem Nocturne" by the Viscounts that featured a throaty saxophone and an electric guitar with lots of tremolo. I also liked the song "Moon River" as sung by Nat King Cole, because it reminded me of the pond beside my childhood home and the warm spring nights when frogs were singing. I took the title of the song "The Night Has a Thousand Eyes" by Bobby Vee quite literally, and imagined the darkness filled with millions of eyes, like indistinct, interlocking peacock eyes. Of course, one of the greatest night anthems is "Tonight," from the score of *West Side Story*. It is a delicious anticipation of the night, echoing Percy Bysshe Shelley when he wrote: "I ask of thee, beloved Night— / Swift be thine approaching flight, / Come soon, soon!"

Musicians have always written about the night, and during the explosion of popular music during the last century, night was praised in dozens of famous songs. Titles like "Stardust," "Shine on Harvest Moon," "I Could Have Danced All Night," and "Moondance" are just a few that represent the deep, musical fascination that night has held, and continues to hold, for us. The blues also embraced night, as has jazz, which, in the last half of the twentieth century, has become an almost exclusively urban-nocturnal form, as exemplified in Thelonious Monk's "'Round Midnight." In the film biography of the same title, a series of urban, nocturnal tableaux evoke the incestuous relationship between jazz and night. Film, like music, is a time-based art, and film has also had a tremendous influence not only on how we see the night, but also on how we experience it.

Film Noir

Earlier in the book we surveyed horror films, which certainly represent the Hecatian aspects of darkness, though there are many films that have depicted twilight more benignly. Unique and cinematographically excellent night scenes can be viewed in films as diverse as *Rebel Without a Cause*, *Black Orpheus*, and *Night of the Shooting Stars*. But if any genre of film epitomized night and darkness it was film noir. Visually slick and aggressively stylized, film noir had villains, heroes, and femme fatales who inhabited an eternal night that will forever be a part of our cultural legacy. The term *film noir* was coined by two French critics who noticed that postwar American crime and detective films had an eerily dark, nocturnal look to them—most of their scenes were set in twilight and shadows.

Film noir was characterized by innovative techniques of lighting that became part of its style—interiors were defined by expressionistic lighting and multiple shadows. As well, disorienting, tilted camera angles were used, and for a purely decorative touch the ever-present tendrils of cigarette smoke were displayed rising in languorous arabesques from the fingers of lethal blonds in the ubiquitous twilight. Cigarette smoke became a sort of cinematic signature of the genre. Venetian blinds were also ubiquitous, their shadows slicing walls into bands of black and white. The cynical narratives of noir unfolded in the twilit interiors of cheap hotel rooms and apartments, on rain-slicked evening streets and in dingy offices. Night à la noir was a glamorous, violent medium that bred danger and romance.

But film noir was not without celluloid and literary precedents. Among its foremost literary predecessors were the "hard-boiled" detective novels of Raymond Chandler and Dashiell Hammett. Noir's cinematic influences came from four prior genres: the American gangster and crime movies of the thirties, German expressionism from the Roaring Twenties, French poetic realism of the thirties, and Italian neo-realism. It was the influence of Italian neo-realism that led to the noir focus on location shots, subtle characterizations, and later, to documentary, style narration, which was extensively relied upon in the TV series *Dragnet*, for example. If the film medium had been in its infancy during the twenties and thirties, it came of age in the early forties. Cinema became worldly with noir; it lost its innocence and began to master new technological advances, mostly in lighting and film quality. Sound had also reached new levels of sophistication. But most important, film noir was cinema's first moment of self-consciousness, where the artificial night of the darkened movie theater spilled onto the screen and began to change the content of the movies themselves. Film noir depicted a world of interminable night populated by world-weary detectives and two-bit losers. The titles say it all: *The Big Sleep* (1946), *Night Has a Thousand Eyes* (1948), *He Walked by Night* (1949), and the classic *Night and the City* (1951). These movies were successive installments in a grand theatric darkness, but film noir also depicted our shadow side, where instincts and desires were released in the night to dance their strange, sometimes wicked choreographies. And there were moments when film noir and horror overlapped—the blood-curdling female screams that accompanied the discovery of a murdered body were identical to those that signaled the attack of a vampire.

Film noir broke a few moral barriers at the time it first emerged. The nighthawks who peopled these films were as dark as the moody cinema screens they dominated. Some noir characters were sexually perverse, or at least deviant by allusion, and many were sociopathic. Film noir was also the first psychologically informed Hollywood genre—Freudian ideas about the unconscious saturated character motivation. A preoccupation with good and evil, with guilt and remorse, underlaid the moral conflict in film noir, where often as not a good person wrestled either with his or her own morality, or the immorality of others. The emotional landscape that these characters inhabited was one of desolation, alienation, cynicism, ambiguity, corruption, guilt, and paranoia. The men were private eyes, gangsters, villains, cops, crooks, thugs, blackmailers, and yes-men—they were sardonic or conflicted, wrestling with brooding disillusionments and insecurities. The women were one of two types: wholesome, trustworthy, monogamous, and reliable or predatory femme fatales, at once gorgeous, mysterious, manipulative, double-crossing, and desperate.

The classic film noir plot was usually a nocturnal tragedy that involved doomed lovers or individuals who fell from grace. These characters were driven by obsessions and seemed to have little insight into their actions, like Lana Turner and John Garfield in Tay

Garnett's night-soaked *The Postman Always Rings Twice* (1946). In another sense it is as if film noir took place entirely in the shadowy world of the unconscious, in the twilight zone of repressed deviancies and obsessions. Night became a proscenium for dark urges.

Film noir didn't monopolize cinematic depictions of night in movies, and as we've seen, many other genres took a more romantic view of darkness. In the visual media, film is a much more recent addition to the spectrum of the arts, and painting, like photography, captures moments rather than sequences. There are many excellent photographs of night and to list them here would be exhaustive, though a recent collection of urban night photographs by Lynn Saville is worth noting for their haunting realism. It is painting rather, that has the longest history of capturing night, and it is painting that is gesturally imbued with all the movement and intelligence that the artist brings to the work.

Painting Night

For my part I know nothing with any certainty but the sight of the stars makes me dream.

VINCENT VAN GOGH

The earliest artistic depictions of night, or at least of the moon and stars, are found on painted Egyptian limestone reliefs more than three thousand years old. There is one particular image that I've always found both erotic and poetic, of the sky goddess Nut, naked, with her long, thin body arched over the earth. She is looking down with a serene gaze, her small breasts pointing to the earth and only her outstretched fingers and toes touching either horizon. The earth is her brother and lover, Geb, who is sometimes depicted as ithyphallic and straining up toward her as she looks down. In the images of her I like the best she is representing the sky at night and her body is covered with stars.

There are pictographs from North America, Africa, and Australia that have been interpreted as depictions of the moon, or series of moons, perhaps as a way of representing the passage of time, though dating these has been problematic. Similar to these are the lovely and well-preserved images of a crescent moon, a full moon, and the constellation Pleiades, applied in gold leaf on a disk uncovered by archaeologists in Nebra, Germany. The Nebra disk has been dated as being thirty-six hundred years old, making it contemporary with the earliest Egyptian reliefs. Certainly the Greeks depicted Selene, the moon goddess, on their painted pottery a thousand years later. But the first real paintings of night, or at least nighttime activities, come in the form of frescoes from Pompeii and Herculaneum, two southern Italian resort towns of the Augustan age that were famously destroyed (and paradoxically preserved) during the eruption of Mount Vesuvius on a hot sunny day in August A.D. 79.

Many erotic frescoes were found in Pompeii, among which was one that depicted a husband and wife engaged in sex. They are naked, though she is still wearing her *strophium* (brassiere), and are making love by lamplight in their domus. In another fresco from Hercula-neum's gymnasium, a young man is drinking wine in a dining room on a warm night after dinner. He has his arm around the shoulders of a young woman, and both are naked from the waist up. Other frescoes depict evening entertainments, and some even have dialogues written in Latin superscripts above the speakers' heads.

But nighttime activities are not depictions of night itself. What might be the only true representation of night from Pompeii is an inlaid marble wall panel found in a private residence. It is a Dionysian scene showing a satyr and an ecstatic maenad. She is dancing beside a small shrine with a lit torch in her hand. The black slate into which these figures are inlaid is meant to be the night itself, congealed into stone.

The tradition of image making, which had formerly been religious, had become secularized in the high cultures of Greece and Rome, and it became devotional once again after the fall of Rome and the rise of Christianity and Islam. Images of religious significance, primarily as manifested in "illuminated" manuscripts both Christian and Muslim, predominate through the Dark Ages until the Renaissance.

Islam has always had a close relationship to night, dating its religious observances by the lunar calendar and celebrating several nocturnal events from Muhammad's life. There is a jewel-like painting from an illuminated Persian manuscript dating about 1540 (though it was painted in a style that predates the Renaissance), showing Muhammad riding a chimerical horse with a woman's head during his ascent to heaven from Jerusalem. He is enclosed by a flaming aura of gold as he rises above the clouds into a deep, Persian-blue night sky filled with golden stars. On his journey to heaven Muhammad is surrounded by angels with brilliantly colored wings. Both the clouds and the flames show a Chinese stylistic influence, echoing the Qur'an's edict to "seek knowledge, even unto China."

Just before the Renaissance, religious art began to assume a more secular appearance, particularly in the attention it gave to pictorial accuracy. A large, devotional painting of the Adoration of the Magi by Gentile da Fabriano was completed in 1423 and consists of a large painting of the Adoration of the Magi surrounded by smaller paintings inset into the ornate, gilded frame. In the lower left-hand corner is a night scene of the Nativity with the infant Jesus, the Virgin Mary, some cattle, and a few onlookers. Da Fabriano's Nativity has a homey, rural feel to it. The source of the light, however, is the most extraordinary and innovative aspect of this painting. The onlookers' faces and upper bodies are illuminated by what looks like firelight, but there is no fire. By looking at the angle of the shadows you realize that this light can only be emanating from the baby Jesus, though his body does not glow. It is at once natural and unnatural.

(This technique of using the baby Jesus to provide illumination became a standard technique.) Above this scene, over the shadowy mountains behind the nativity, is a star-sprent, deep blue sky that stretches from horizon to horizon. Da Fabriano was an observant artist and his use of the costly pigment lapis lazuli, made by grinding down the gemstone into dust, ensured that the blue of the night sky was accurate as well as precious.

During the mid Renaissance artists began to abandon the bright colors of the early Renaissance for more somber, blended tones. As a result the figures in many Renaissance paintings seem to emerge from a twilit world of shadows. This is particularly true of the paintings of the High Renaissance painters like Leonardo da Vinci and Titian in Italy and Albrecht Dürer and Pieter Brueghel the Elder in northern Europe. This twilight trend continued through the Mannerism period and was also incorporated by the Baroque artists El Greco and Tintoretto. One masterpiece from the Northern Renaissance, painted by Mathias Grünewald, captures the essence of the boreal night of the High Northern Renaissance.

Grünewald's *Isenheim Altarpiece*, a series of panels depicting the life, crucifixion, and resurrection of Christ, is considered one of his finest works. Here Grünewald combined the bright, saturated colors of the early Italian Renaissance with two spectacular night scenes. In the first panel, painted on the outside-facing part of the altarpiece, Grünewald portrayed the black night of the crucifixion, which, according to the New Testament, spread over the land when Jesus died on the cross. Starless and desolate, it is a night of calamity, of supernatural remorse and loss.

When the *Isenheim Altarpiece* is opened, the inside panels reveal the life of Jesus, reading from left to right, ending with a final panel representing Christ's resurrection at night, a joyous night that is clear and star-filled. The painting shows Christ just moments after he has flung aside the huge stone blocking his tomb, as he rises into the air with his body in a position of mystic beatitude. He stares out of the canvas at the viewer, his arms upraised in holy rapture while the stigmata on his hands and feet shine like flames. His body is preternaturally pale, animated by divine force, and he rises amid a bright, circular halo that radiates all around his upper body, like a spherical aurora borealis as he ascends into a deep blue night filled with hundreds of stars.

As the Renaissance gave way to the Baroque period, around 1600, the dark, somber painting style of the Renaissance continued with painters like Velázquez, Caravaggio, and Rembrandt. But departing from this tradition was Adam Elsheimer's *Flight into Egypt* (1609), which was perhaps the first completely realistic painting of night in which the majority of the canvas is devoted to the sky. The painting, set in the north European countryside, shows a silhouette of dark trees diminishing into perspective from left to right. Just where the trees reach the horizon, an opalescent full moon rises among a few clouds, its image reflected in a lake below. There are two other sources of light in the painting, both of which emerge out of the darkness of the forest in the lower part of the painting. The central light is the torch that Joseph carries. It illuminates, dimly, the donkey with Mary carrying the baby Jesus. To their left, and in the lower corner of the painting, two shepherds watch a fire whose sparks rise up above the trees and blend into the stars that themselves float in the soft blue summer night. The stars are beautifully accurate, and you can even recognize constellations as well as the Milky Way. Overall the painting conveys a peaceful, wondrous enchantment that reinforces the quiet spirituality of its subject.

Rembrandt's *The Night Watch* (1642) is reputedly one of the most famous depictions of night, and his extraordinary treatment of illuminated, nocturnal human figures is one of the highest expressions of Baroque technique. (I say *reputedly* because some art historians claim that *The Night Watch* was not the original title and that it was simply its varnish darkening over the centuries that gave the painting its twilit cast.) It certainly is an unearthly night that Rembrandt has painted, with a strangely interior feel. The group assembled for this portrait, dressed in the flamboyant Baroque style of the Dutch military, seems to be in a large, dark room, rather than outside. The twilit ambience of *The Night Watch* is similar to the quality of night on city streets or on porches and patios at night. With electric lighting you cannot see night's limitlessness, and it becomes possible to imagine that you are actually in some large room whose walls you cannot see. Rembrandt has somehow anticipated this artificial lighting that would not become a reality for more than two hundred years, for unlike the natural light of Da Fabriano, Rembrandt throws light on particular individuals of the group, two figures accented with yellow clothing, giving the scene a spotlit feel.

After the Baroque period, toward the end of the eighteenth century, the subjects of painting became much liberated from their classical predecessors. The fascination with horror and terror that fostered the Romantic Gothics, for example, was well represented in the work of John Henry Fuseli. His painting *The Nightmare*, completed in 1790, combined terror and hallucination in a painting that foreshadowed surrealism by one hundred and fifty years. In the painting a white-robed sleeping woman, whose arms and head are draped off the end of a bed on the left-hand side of the painting, has a malign, grinning incubus sitting on her chest, while behind the incubus, its head protruding through an open curtain, is the upper body of a white horse, literally a "night mare" whose eyes are eerily white and opaque. The somber gray and brown tones of the painting contribute to the claustrophobic, nightmarish atmosphere of a truly unsettling, psychologically disturbing scene that continues to be widely reproduced. Just as Fuseli anticipated the Gothic genre of the nineteenth century, so did Francisco Goya anticipate the gritty realism of the twentieth century.

There are nights of anguish and torment, and artists have never shrunk from the task of depicting some of the darkest chapters in human affairs. The theme of night and horror that Fuseli introduced was picked up a few decades later by Goya. His painting *The Third of May, 1808* is one of the most dramatic portrayals of a night of despair and horror ever produced. It shows, in graphic detail, a mass execution on a dark, black night. The central scene emerges out of the darkness—to the right soldiers lean forward with their rifles pointed at the prisoners, who, on the left, are illuminated coldly by a single box lamp. The prisoners are pleading for their lives on the fallen, bloody corpses of their predecessors. Only *Guernica*, by Pablo Picasso, comes close to Goya's graphic depiction of the bloody terror of war. Above the execution scene the night sky is pitch black, devoid of any blue, as if this massacre were taking place at the bottom of an endless pit of darkness. The dark night of the soul spreads its empty, bleak dominion over a kingdom of despair.

In England, at the same time as Goya was working, the Pre-Raphaelite painters were embarking upon their allegorical, neomannerist romantic movement. Artists like Dante Gabriel Rossetti and William Holman Hunt were depicting classical, romantic themes. One of them, Edward Burne-Jones, painted several allegorical works about night. Two of these were paintings of a woman floating above a landscape; the first was called *Evening Star* while the second was simply called *Night*. Painted in 1870, *Night* combines watercolor and body color on canvas. A woman floats in the sky, high above an indistinct landscape. She personifies Night in the Pre-Raphaelitic tradition of neoclassical allegory, and her long flowing robes cover most of her body, which is turned slightly away from the viewer so that her face is hidden. The night sky is blue and filled with stars. She holds her hands up in the air in front of her and to the right, as if she were undulating them in the midst of ritualistic, airborne dance.

James Abbott McNeill Whistler was an American painter associated with the Pre-Raphaelites, though his work was so individualistic his inclusion within the movement is tenuous at best. His impressionistic *Nocturne in Black and Gold: The Falling Rocket* (1874) illustrates a magical, poetic darkness with a decidedly un-Pre-Raphaelite style that seemed to anticipate the work of Claude Monet rather than look back at neoclassicism. Because of its abstraction, it garnered considerable critical scorn at the time. By today's standards it hardly seems avant-garde—instead it looks almost whimsical. A trail of embers or sparks meanders down through the night sky above a nocturnal vista of tall, dark trees, water, and distant buildings. Whistler's *Nocturne* depicts a symphonic, luscious summer night filled with luminous points of light. His canvas is awash in deep greens and blues, lending the night sky an aquatic, underwater look. The glowing lights of the buildings, almost indistinguishable from the rocket embers, add a dreamy depth. This painting seems like the visual equivalent of one of Satie's or Debussy's tone poems.

Whistler described the mood of twilit landscapes, perhaps with his *Nocturne* in mind, during a lecture he delivered in England. Oscar Wilde happened to be in the audience and in his subsequent review of Whistler's lecture in the *Pall Mall Gazette* he recounted how Whistler "spoke of the artistic value of dim dawns and dusks, when the mean facts of life are lost in exquisite and evanescent effects, when common things are touched with mystery and transfigured with beauty; when the warehouses become as palaces, and the tall chimneys of the factory seem like campaniles in the silver air."

By the time Vincent van Gogh painted *Sidewalk Café at Night*, in 1888, electric lighting had spread to the smaller towns. His painting of a brightly lit café patio on a cobbled street in the southern resort town of Arles is sensual and inviting. You can almost hear the soft laughter of the patrons lingering on the patio, the clinking of glasses and cutlery from inside the restaurant. A waiter is serving some guests on the patio and all are bathed in a luscious yellow light. Van Gogh believed that every color had a symbolic meaning and he equated yellow with love. There is certainly love in this warm, delightful citron that spills out over the street, contrasting with the blue darkness surrounding it. But it is van Gogh's blue night sky that elevates this painting into a celestial realm; for blue, according to van Gogh's symbolism, meant divinity, and this painting is saturated with it. Above the indigo street, above the blue-shrouded buildings inset with a handful of orange windows, is an ultramarine night sky filled with stars. But this night is deeper than any he had yet painted. The stars are huge, white-hot discs with yellow haloes that make them seem like lamps hanging in the sky. Van Gogh used haloes sparingly because they symbolized "something of the eternal," as he once wrote.

A year later, as he began his slide into terminal psychosis, van Gogh produced another night scene, but this one transcended everything he had done before. Through excruciating, unmediated torment, his soul had finally arrived at a narrow divide between heaven and hell, where, teetering, he could look into both infinities with uncompromising clarity. Something ethereal, pure, and spiritual about night and stars was calling to him. As he wrote his brother: "That does not keep me from having a terrible need of—shall I say the word—religion. Then I go out at night to paint the stars." The ultimate expression of his enchantment by the warm nights in the countryside around Arles is called *The Starry Night*, painted in 1899 near a small town called St. Rémy.

It is clear that van Gogh's visionary powers are even more magnified in this painting than in his previous nocturne, *Sidewalk Café at Night*. The dark sky is filled with stars whose haloes are even larger than those he painted a year earlier—they have become spiraling mystic orbs—and the peacock-blue twilight is writhing with flowing patterns, with vortices of color and wind. This invisible force which pervades van Gogh's night affects even the trees, which seem to curl and bend like dark flames as they spire into a sky in which the entire cosmos seems present with mesmeric power. The artist's brushstrokes are laid on like tiles of pigment, and the painting is a mosaic of color and movement. If it seemed that you could walk into the twilight of *Sidewalk Café at Night*, then *The Starry Night* opens up and swallows you.

At the same time that van Gogh was in Arles, Henri Rousseau, France's most celebrated naïve artist, was painting his spellbinding canvases in Paris. Rousseau was an idiot savant of sorts, and post-humously became a favorite of the surrealists. He was famously innocent yet, at the same time, he was capable of producing strangely compelling dreamscapes with unmistakable authority. His painting often included tropical forests replete with exotic flowers and animals, and several of his best works were set at night.

One of these, *The Sleeping Gypsy*, painted in 1897, is of a North African man lying asleep in the desert. The landscape around the sleeper is desolate and haunting—low distant mountains and a river—and embodies the essence of unconsciousness, night, sleep, and dreams. The sleeper is wearing a striped robe. Beside him lies a lute and a vase. The central focus of the painting, however, is behind the sleeper. There, startling, in hallucinatory clarity, is a lion. The lion is standing in profile, facing right, with his nose close to the ground as if scenting the sleeper's spoor. His mane is blowing in the night wind, and his yellow eye is at once menacing and hypnotic. Above the lion and sleeping man, the warm night sky is a pale blue, saturated with the light from a full moon. A few stars glow beside the moon. This night is filled with lyric mystery, and seems more like a lucid dream. No wonder Rousseau was also a favorite painter of the psychoanalysts.

Another extraordinary night painting of Rousseau's, and possibly his most reproduced work, if not his most famous, is called *The Dream*, and was painted in 1910. Here, in the moonlit foliage of a jungle night, the dreamy quality of his earlier work is intensified to produce something completely surreal. Set in the midst of the lush vegetation is a red velvet couch on which a beautiful nude woman reclines, bathed in the light of the full moon. To her right a snake charmer

with hypnotic eyes plays upon his flute. Perhaps it is his music that
has conjured the jungle animals gathered around him. All seem
bewitched in the magic darkness. The details of the foliage, the
extraordinary symmetry of the flowers and plants, combine to transform this preternatural night into a visionary theater of magic.

The pastoral scenes of night that characterized many classical
paintings gave way, in the twentieth century, to cityscapes. Artists
began to treat cities as the new landscape, and Edward Hopper was one
of the foremost urban painters. Hopper once said his favorite European
artist was Rembrandt, and that his favorite painting by Rembrandt
was *The Night Watch*. It's no wonder then, that Hopper's most famous
painting, *Nighthawks*, painted in 1942, was a direct homage to *The
Night Watch*. Not only are both paintings set in an urban night, but
both also portray people dressed in contemporary clothing and allude,
in a subtle way, to relationships between these people. Hopper's
Nighthawks is a masterpiece of twentieth-century realism; it is dominated by the view of a downtown, late-night diner with a man and
woman sitting at a long, wraparound, forties-style counter. A waiter
stands at another section of the counter, as if the painting were a
documentary photograph of a moment on an average night.

The palette of this work is remarkable. The restaurant, illuminated
with a green light, looks as if it were an aquarium enclosed in large,
plate-glass windows. Outside, on the sidewalk in front of the
windows, an anonymous man walks by, bathed in the green light
flooding from the diner. While the verdant circle of illumination cast
by the diner dominates the scene, in the darkness of the street across
from the diner, faint, purple-colored windows glow in the city night.
There is no mistaking the mood of a city late at night: the couple
relaxing after an evening on the town, or perhaps having a midnight
rendezvous, the stranger out walking late, and the waiter, working
the night shift in a city that never sleeps.

Although his technique was not as accomplished as Salvador Dalí's, or his output as varied and prodigious as that of Max Ernst, René Magritte was possibly the greatest of the surrealist artists. His painting style had a practical, masterful earthiness that clearly and simply conveyed his strange and unsettling compositions. Magritte produced several night scenes, most notably the series titled *The Dominion of Light*, a suite of paintings executed in the early 1950s that depicted nighttime street scenes—including houses, streetlights, and dark trees—with daytime skies above them, that are possibly his most reproduced works.

One of the most famous of these, *The Enchanted Domain*, portrays a typical Brussels residential street at night. The boulevard in front of the simple three- and four-story houses is deserted, illuminated by a single streetlamp. A few windows are glowing cozily in the twilight, though most others are dark, indicating a late hour. The buildings are all gray and behind them loom some dark trees. But startlingly, above the dark silhouette of the trees and roofs, is a bright blue sky, dotted with puffy, fairweather cumulus clouds. It is this juxtaposition, strangely appropriate, that creates both the tension and the poetry of the painting's vaguely portentous atmosphere. It is charged with symbolic meaning, yet the overall mood of *The Enchanted Domain* is quiescent and idyllic, filled with an overriding sense of calm wisdom pervading the mysterious night scene paradoxically placed beneath the bright blue sky of a beautiful summer day. The coexistence of night and day in this painting produces a visual conundrum equivalent to an optical illusion in the viewer. My own, somewhat involuntary response is to turn the painting into a morning scene. After all, there are times where the streetlights are still on and yet the eastern sky is glowing with the colors of dawn. Perhaps it is at dawn, rather than at sunset, that night and day are most commingled, that dwindling pockets of night linger in dark pools as the morning begins to break.

14

NIGHT'S LAST STAND

Awake! for Morning in the Bowl of Night
Has flung the Stone that puts the Stars to flight:
And Lo! the Hunter of the East has caught
The Sultan's Turret in a Noose of Light.

EDWARD FITZGERALD

AS MYSTERIOUSLY AND incrementally as night arrived, it begins to fade away. The eastern sky lightens and the stars grow pale and disappear one by one. Soon, all that will be left of the evening will be forensic clues—a garbage can tipped over by raccoons, empty wine bottles and glasses on a patio table, diamonds of safety glass adorning the gutter of a quiet side street, a splash of feathers and blood on a snow-covered field, a jewelry box on a bedside table.

These are the last minutes of night, though deep within forests, in mountain valleys and city alleys, it is still dark—night's last stand. It is here, in these surviving pools of twilight, that darkness still reigns, like an exile or a memory, like the magic night in little Melissa's box. But eventually, even these succumb to the irresistible flood of light from the sun. Except some nights are strong, and they survive the dawn in our memory, lingering through the day like the remnants of a dream, and the mood of special nights can linger even longer. Perhaps one of the most lyrical reminiscences of night is contained in Marguerite Duras's *The Lover*, where the novel's protagonist recalls a series of extraordinary tropical nights she experienced in her childhood:

Sometimes, it was in Vinh Long, when my mother was sad she'd order the gig
and we'd drive out into the country to see the night as it was in the dry season. I
had that good fortune—those nights, that mother. The light fell from the sky in
cataracts of pure transparency, in torrents of silence and immobility. The air
was blue, you could hold it in your hand. Blue. The sky was the continual
throbbing of the brilliance of the light. The night lit up everything, all the
country on either bank of the river as far as the eye could reach. Every night was
different, each one had a name as long as it lasted. Their sound was that of the
dogs, the country dogs baying at mystery. They answered one another from
village to village, until the time and space of the night were utterly consumed.

Duras captures the astonishing depth of the tropical night, for
twilight in the tropics is indeed enchanted. I too have experienced
evenings that were perhaps equally enchanted, though many of the
evenings I remember vividly were not just in the tropics. These were
exceptional nights that enlarged my vision of what twilight could be,
nights of unbelievable clarity and magic. I can remember stars over
the Hawaiian Islands so large and luminous they startled me, like a
celestial city in the sky. One evening, years ago, I slept in a farmer's
field near Stonehenge, England, on a midsummer's eve under a velvet
night sky and saw stars that seemed to glow from the tips of invisible,
purple stalactites, hanging from the vault of heaven. I have watched a
full moon rise above Mayan ruins, still warm from the Mexican sun,
beside a bottomless cenote. I have seen the aurora borealis dance
electric over a lake in northern Canada, dazzling with shimmering,
ever-changing pulses across the entire summer night sky. I've
watched bats replace swallows as the mild Neapolitan sunset gave
way to twilight near the ruins of Pompeii.

One moonlit summer night in northern Ontario, I paddled a canoe
on a lake so still it was like a liquid mirror, star filled and deep as the
night itself. I've seen clouds lit pink by city light, riding a night sky
as iridescent blue as kingfisher's feathers. I've been surprised by large

full moons, red as burgundy, rising hugely behind a constellated skyline of skyscrapers. In the Lesser Antilles I have walked along beaches where the moonlight seemed as bright as day, and the shadows of the coconut palms were splashed over the white crescent of the beach like inky starfish. These reminiscences of night remain, even here at dawn, and they flavor the day like a subtle spice.

First Light: The Physics of Dawn

The morning stars sang together, and all the sons of God shouted for joy.

JOB 38:7

As the sun approaches the horizon from beneath, the birdcalls that began sporadically half an hour ago become more frequent. Birds love the dawn. Even in the city they break into a sunrise reveille, but in the country the air rings with their calls, reaching a peak as the first sunlight ignites the tops of the trees with an orange corona. There is something insistent, almost digital about their songs. Considering that most paleontologists now believe that birds represent the descendants of winged dinosaurs, the dawn chorus takes on a slightly different tone, at least to my ears. Are these the echoes, perhaps even the refinements, of the calls with which dinosaurs greeted the dawn millions of years ago?

Now, as the edge of the sun becomes visible, the sky incandesces with pinks and reds, and, like the sunset the evening before, there are a few orange-tinged clouds against a yellow sky directly above the sun—but is dawn simply the reverse of dusk? Are sunrises the mirror image of sunsets? I've often speculated that you could take photographs of sunsets and claim that they were sunrises and no one would be the wiser. Yet I think most of us believe we can distinguish dawn from sunset; after all, it seems natural to presume that there must be some difference. And there is.

Sunrise and sunset take place under different atmospheric conditions. At sunset the atmosphere holds many more fine, airborne particles stirred up by the activities of the day—dust carried aloft by convection currents, vegetation aerosols, and plant pollens, as well as smog from cities and factories. As we know, the more airborne particles there are in the atmosphere, the redder the light that passes through them. Also, because the atmosphere is generally warmer at sunset, the cloud formations are often more elaborate, creating the conditions for richer, more varied displays at sundown rather than dawn.

That is not to say that dawns cannot be as fantastic and beautiful as sunsets. I have seen dawns with colorful long, horizontal banks of flaming red clouds against the pale blue sky. But generally speaking, dawns are clearer and more transparent than sunsets. Day is so powerful, so iconic and illuminated, that night fades quickly in its thrall, as though twilight were a dream, a quirky memory.

Preceded by an immense star, almost unearthly in its effusion of white splendor, with two or three long unequal spoke-rays of diamond radiance, shedding down through the fresh morning air below—an hour of this, and then the sunrise.

WALT WHITMAN, JULY 23, 1878

Now, as the purple ocean of twilight subsides on the western horizon, the night that was will never be again. But there will be another night, and another after that, time enough to sing night's praises. For when night falls again, I will drink it in—I will inhale its fragrance, and my skin will thrill to its touch. The spectacle of the stars continues to amaze me and my soul still opens when I hear the victorious cry of the nighthawk from the high, blue vaults above the city—its mysterious nightsong somehow echoing the refrain of summer darkness. I love all the inhabitants of twilight, from glistening limousines to velvety leopards. Night is my country,

my lover, and after sunset I fill with a kind of restless euphoria—
twilight's power flows through my body. I feel lighter, stronger,
quicker, more awake. If my day has been dreary, sunset brings respite.
After all, the same cosmic night in which the galaxies swirl is inside
all substance, inside us, an eternal, brilliant night within and beyond
all matter, for all time.

ACKNOWLEDGMENTS

I extend my initial thanks to Iris Tupholme at HarperCollins Canada for her continuing support and assistance. I also thank Nicole Langlois, as well as managing editor Noelle Zitzer, who oversaw production. Thanks go to Chris Bucci, Ian Murray and the indefatigable Rob Firing. At Bloomsbury USA, I thank Karen Rinaldi for her faith and my editor Gillian Blake for her thoroughness, as well as Amanda Katz and Dena Rosenberg for their assistance and optimism. Thanks go to Alexandra Pringle of Bloomsbury UK for her confidence and to my editor Bill Swainson. Barbara Gowdy also provided editorial advice.

My research was assisted by many, including William Torode of the National Speleological Society, Huntsville, Alabama; Dr. Pat Hanly, director of the Sleep Laboratory at St. Michael's Hospital; Edmund Villamere of the York Regional Police; Joe Mihevc, councillor for the City of Toronto; Lisa Zaritzky; Julie Enfield; Tim Rostrum; Deborah Windsor; Harold Hosein at CityTV; Bruce Bowker; Margaret Atwood; and Jane Urquhart. Grateful acknowledgment is also made to the authors whose work is cited in the book and listed in its bibliography.

In addition, I am indebted to my agent, Bruce Westwood, of Westwood Creative Artists, who proferred the germinal idea for this book as well as providing the time to write it. Finally, I thank Natasha Daneman and Nicole Winstanley, also of Westwood Creative Artists, for all their assistance.

BIBLIOGRAPHY

Abbott, Elizabeth. *A History of Mistresses*. HarperCollins, 2003.

Adams, Peter. *Moon, Mars and Meterorites*. Cambridge University Press, 1986.

Alter, Dinsmore. *Pictorial Guide to the Moon*. Thomas Y. Cromwell, 1973.

Alvarez, A. *Night: An Exploration of Night Life, Night Language, Sleep and Dreams*. Jonathan Cape, 1995.

Apollonio, Umbro, ed. *Futurist Manifestos*. Thames and Hudson, 1973.

Aronson, Alex. *Psyche and Symbol in Shakespeare*. Indiana University Press, 1972.

Barrow, John D., and Frank J. Tipler. *The Anthropic Cosmological Principle*. Oxford University Press, 1986.

Borror, Donald J., and Richard E. White. *A Field Guide to the Insects*. Houghton Mifflin, 1970.

Bulfinch, Thomas. *Bulfinch's Mythology*. Doubleday, 1968.

Carey, John. *The Faber Book of Science*. Faber and Faber, 1995.

Chartrand, Mark. *Collins Guide to the Night Sky*. HarperCollins, 1995.

Cunliffe, Barry. *Rome and Her Empire*. McGraw-Hill, 1978.

Dement, William. *The Promise of Sleep*. Dell, 2000.

Dennett, Daniel Clement. *Consciousness Explained*. Little, Brown, 1991.

Dickinson, Terence. *The Universe and Beyond*. Camden House, 1986.

Dimont, Max I. *Jews, God and History*. Signet Books, 1962.

Duras, Marguerite. *The Lover*. Translated by Barbara Bray. HarperPerennial, 1992.

Durrell, Gerald. *My Family and Other Animals*. Rupert Hart-Davis, 1956.

Epstein, Isidore. *Judaism*. Penguin, 1959.

Ernst, Max. *Une Semaine De Bonté*. Dover, 1976.

Fenton, M. Brock. *The Bat*. Key Porter, 1998.

Fenton, M. Brock. *Bats*. Fitzhenry & Whiteside, 2001.

Ferris, Timothy. *Seeing in the Dark: How Backyard Stargazers Are Probing Deep Space and Guarding Earth from Interplanetary Peril*. Simon & Schuster, 2002.

Flaubert, Gustave. *Madame Bovary: The Story of a Provincial Life*. Translated by Alan Russell. Penguin, 1950.

Freud, Sigmund. *The Interpretation of Dreams. The Basic Writings of Sigmund Freud*. The Modern Library, 1966.

Freud, Sigmund. *Three Contributions to the Theory of Sex. The Basic Writings of Sigmund Freud*. The Modern Library, 1966.

García Márquez, Gabriel. *One Hundred Years of Solitude*. Everyman's Library, 1995.

The Globe and Mail. "Blinded by the Light," January 12, 2002.

Hall, Calvin S. *The Meaning of Dreams*. McGraw-Hill, 1966.

Hardy, Thomas. *Far from the Madding Crowd.* Wordsworth Editions, 1993.

Hervey de Saint-Denys. *Dreams and How to Guide Them.* Translated by Nicholas Fry. Duckworth, 1982.

Hobson, J. Allan. *The Dreaming Brain.* Penguin, 1990.

Holland, W. J. *The Moth Book: A Guide to the Moths of North America.* Doubleday, Page and Company, 1903.

Homer. *Iliad.* Translated by Richmond Lattimore. University of Chicago Press, 1951.

Honegger, Gottfried, and Peter van de Kamp. *Space.* Dell, 1962.

Janson, H. W. *History of Art.* Harry N. Abrams, 1995.

Jones, Ernest. *Sigmund Freud: Life and Works.* Hogarth Press, 1957.

Jung, Carl. G. *The Collected Works, Dreams.* Routledge, London; Princeton University Press, Bollingen Series, Volumes 4, 8, 12, 16.

Knudson Kinishi, Eric I. "The Hearing of the Barn Owl." *Scientific American,* April 1993.

Kortlandt, Adriaan. "Chimpanzees in the Wild." *Scientific American,* May 1962.

Linsenmaier, Walter. *Insects of the World.* McGraw-Hill, 1972.

Longfellow, Henry W. *Poems and Selections.* H. M. Caldwell Co., 1909.

Mansfield, Katherine. *The Collected Stories of Katherine Mansfield.* Penguin, 1985.

Melbin, Murray. *Night as Frontier: Colonizing the World After Dark.* Collier Macmillan, 1987.

Nagel, Thomas. *Mortal Questions.* Cambridge University Press, 1979.

Nathan, Peter. *The Nervous System.* Oxford University Press, 1983.

Nietzsche, Friedrich. *The Philosophy of Nietzsche.* Random House, 1954.

O'Brien, Tim. *The Things They Carried.* Broadway, 1998.

Persinger, Michael A., and Gyslaine F. Lafrenière. *Space-Time Transients and Unusual Events.* Nelson-Hall, 1977.

Solnit, Rebecca. *Wanderlust: A History of Walking.* Penguin, 2000.

Steel, Duncan. *Marking Time: The Epic Quest to Invent the Perfect Calendar.* John Wiley & Sons, 1997.

Stierlin, Henri. *The Roman Empire.* Taschen, 2002.

Thomas, Lewis. *The Lives of a Cell.* Viking, 1974.

Toman, Rolf. *Baroque: Architecture, Sculpture, Painting.* Köneman, 1998.

Walters, Mark Jerome. *The Dance of Life: Courtship in the Animal Kingdom.* Arbor House, 1998.

Waugh, Alexander. *Time: From Micro-seconds to Millennia—A Search for the Right Time.* Headline, 1999.

Whitehouse, David. *The Moon: A Biography.* Headline, 2001.

Whitman, Walt. *The Portable Walt Whitman.* Viking, 1945.

Wise Brown, Margaret. *Goodnight Moon.* Harper & Row, 1947.

Wright, Lawrence. *Warm and Snug: The History of the Bed.* Routledge & Kegan Paul Limited, 1962.

Young, Edward. *Night Thoughts on Life, Death and Immortality.* R. Dodsley, 1743.

Young, J. Z. *Programs of the Brain.* Oxford University Press, 1978.

INDEX

acetycholine, 146
Adoration of the Magi, 288–289
aircraft, 203
aldosterone, 139–140
Aldrin, Buzz, 242
aliens, 198–201
Amanita muscaria, 195–196
American Civil War, 204
Andromeda, 216–217
anglerfish, 276
animals
   bats, 20, 51–58
   bioluminescence, 44, 68–72
   birds, 58–64, 142, 300
   cats, 47, 48–49, 87, 141, 181–182
   cave dwelling, 270–271, 273
   Eocene period, 267–269
   farm animal mutilations, 199
   fishes, 44–45, 71, 271, 275
   hearing, 60–61
   insects, 55–56, 64–70, 71–72, 141–142, 234, 264
   light pollution and, 103–104
   night monkeys, 48–49
   night vision, 46–51, 60
   sleep and, 141–142
   sounds of, 181–182
   sunsets and, 21
   urban, 62–63, 87
ants, 141–142
apnea, 143, 155, 156, 253
Apollo moon program, 241–244
Arctic Circle, 266–269
Aristotle, 144, 165–166
Artemidorus of Dalis, 166
Artemis. *See* Selene
Artificial Paradise (Baudelaire), 172

artistic depictions of night, 281, 286–297
Aserinsky, Eugene, 144–145
astronomy/astronomers, 106, 223–229
Athena, 58–59
atonia, 148–149, 150, 179
atrocities at night, 202–203
Augustine, 166–167
aurora borealis (northern lights), 244–246, 269
Axel Heiberg, 267–269
Aztecs, 232

banshees, 194
bats, 20, 51–58
Baudelaire, Charles, 172, 201
Baudrillard, Jean, 265
bears, 273
Bermuda fireworm, 71
Bernhardt, Sarah, 116–117
Bertolini, Enzo, 132–134
Betelgeuse, 218
Bible, 9, 163–164, 231
big bang, 11, 42
Big Dipper. *See* Ursa Major
bioluminescence, 44, 68–72
birds, 47, 49–50, 58–64, 142, 300
Blackfoot Indians, 232–233
black holes, 221
Blask, David, 104–105
blues at night, 261–263
Bonaire, 24
bonfire festivals, 119–128
Bourne, William, 226
Bowker, Bruce, 24
Brahe, Tycho, 225
brain waves, 145–146
Bridge on the River Kwai (film), 53–54

A NOTE ON THE AUTHOR

Christopher Dewdney is a poet who has also written extensively about language, culture, and media. The author of three books of nonfiction—*Last Flesh, The Secular Grail*, and *The Immaculate Perception*—as well as eleven books of poetry, he is a three-time nominee for Governor General's Awards and a first-prize winner of the CBC Literary Competition. Dewdney lives in Toronto, Ontario, where he teaches advanced writing at the Glendon Campus of York University.

A NOTE ON THE TYPE

Linotype Garamond Three is based on seventeenth-century copies of Claude Garamond's types, cut by Jean Jannon. This version was designed for American Type Founders in 1917, by Morris Fuller Benton and Thomas Maitland Cleland, and adapted for mechanical composition by Linotype in 1936.